Recent Advances in Motion Analysis

Recent Advances in Motion Analysis

Editors

Francesco Di Nardo
Sandro Fioretti

MDPI • Basel • Beijing • Wuhan • Barcelona • Belgrade • Manchester • Tokyo • Cluj • Tianjin

Editors
Francesco Di Nardo
Department of Information
Engineering, Università
Politecnica delle Marche
Italy

Sandro Fioretti
Department of Information
Engineering, Università
Politecnica delle Marche
Italy

Editorial Office
MDPI
St. Alban-Anlage 66
4052 Basel, Switzerland

This is a reprint of articles from the Special Issue published online in the open access journal *Electronics* (ISSN 2079-9292) (available at: https://www.mdpi.com/journal/electronics/special_issues/motion_analysis).

For citation purposes, cite each article independently as indicated on the article page online and as indicated below:

LastName, A.A.; LastName, B.B.; LastName, C.C. Article Title. *Journal Name* **Year**, *Volume Number*, Page Range.

ISBN 978-3-0365-0438-4 (Hbk)
ISBN 978-3-0365-0439-1 (PDF)

Contents

About the Editors

Francesco Di Nardo is Senior Staff Scientist in the Movement Analysis Lab, Department of Information Engineering Università Politecnica delle Marche, Ancona, Italy. Within the same Department, he is Head of the section "Acquisition Systems and Data Processing". He served as Professor of Medical Informatics in the Bachelor Degree in Biomedical Engineering, Università Politecnica delle Marche, for the academic year 2019/2020. He is currently Associate Editor of the journal *IRBM* (*Innovation and Research in BioMedical Engineering,* Elsevier) and Topic Editor for the journal Electronics (MDPI). In addition, he is the contact person responsible for the region of Marche for the Italian Society of Movement Analysis in Clinics (SIAMOC), Senior Fellow of Interuniversity Centre of Bioengineering of the Human Neuromusculoskeletal System (BOHNES), and a founding member of the association Italian National Group of Bioengineering (GNB). The Università Politecnica delle Marche, where he is currently associated, is also where he completed his master's degree in Electronic Engineering in 2000 and received his Doctor of Philosophy (PhD) in Artificial Intelligence Systems in 2005. His main research activities include the areas of biomedical signal processing (filtering, feature extraction, pattern recognition, time–frequency analysis, application of neural networks to biosignals, statistical gait analysis) and interpretation (physiology, clinics, sport), particularly the acquisition and processing of surface electromyography (EMG) signals to assess muscular function during gait tasks. In this and other biomedical fields, he is author and co-author of more than 130 publications, including full papers in refereed international journals, chapters in international books, and conference papers.

Sandro Fioretti received his Dr Eng degree in Electronic Engineering from the University of Ancona, Italy, in 1979, and diploma for the specialist course in "Engineering of Control Systems and Automatic Computing" from Rome University "La Sapienza" in 1984. Since 2000, he has served as Associate Professor of Bioengineering at Università Politecnica delle Marche, Ancona, Italy. He has participated in various European and National research projects in the field of Rehabilitation Engineering. His main research interests are in movement analysis, stereophotogrammetry for movement analysis (with markers and markerless), linear and nonlinear filtering of biomechanical signals, joint kinematics, modeling and identification of postural control, static and perturbed posturography, telematic applications for movement analysis, rehabilitation engineering, and analysis of biomedical signals.

Preface to "Recent Advances in Motion Analysis"

In recent years, research in the field of motion analysis has contributed to the increasing analysis of motor tasks and activities performed in ambulatory or domestic/work environments. This requires the use of instrumentation that largely differs from that used in classic structured environments such as (clinical) movement analysis laboratories. In particular, the advent of miniaturized sensors, such as inertial sensors with low-cost, wearable, and wireless characteristics, has captured the attention of researchers as evident in the scientific literature of the last decade. Attention has mainly been focused on problems related on how to ameliorate the overall level of accuracy, the proper placement of sensors with regard to which body segment(s), and feature extraction from standardized motor tasks, e.g., f.i., gait, or sit-to-stand or squat. The possibility to monitor generic activities of daily life is nowadays possible as far as the recording of data is concerned. Toward this purpose, we are assisting through the fruitful integration of movement analysis with tools and methods derived from Internet of Things (IoT), automotive, robotics, and gaming contexts along with the availability of low-power consumption and high memory capacity of modern electronic devices. Besides problems of acceptability and usability of the systems, that imply use of the minimum possible number of sensors as well as topics concerning subject privacy, it is the interpretation of the acquired signals that remains a big challenge. The problem of automatic recognition of generic daily activities in long-term monitored signals is an open problem. Valuable contributions are expected from computational intelligence methods such as f.i., artificial neural networks, fuzzy logic rules, and automatic classification methods. In this context, this Special Issue on recent advancements in movement analysis attempts to cover some of the methodological and applicative aspects of modern movement analysis.

Francesco Di Nardo, Sandro Fioretti

Editors

Review

Recent Progress in Sensing and Computing Techniques for Human Activity Recognition and Motion Analysis

Zhaozong Meng, Mingxing Zhang, Changxin Guo, Qirui Fan, Hao Zhang *, Nan Gao and Zonghua Zhang

School of Mechanical Engineering, Hebei University of Technology, Tianjin 300401, China; zhaozong.meng@hebut.edu.cn (Z.M.); mingxing.zhang123@outlook.com (M.Z.); changxinguo1@outlook.com (C.G.); qirui.fan@outlook.com (Q.F.); ngao@hebut.edu.cn (N.G.); zhzhang@hebut.edu.cn (Z.Z.)

* Correspondence: zhanghao@hebut.edu.cn; Tel.: +86-1382-1425-017

Received: 28 July 2020; Accepted: 18 August 2020; Published: 21 August 2020

Abstract: The recent scientific and technical advances in Internet of Things (IoT) based pervasive sensing and computing have created opportunities for the continuous monitoring of human activities for different purposes. The topic of human activity recognition (HAR) and motion analysis, due to its potentiality in human–machine interaction (HMI), medical care, sports analysis, physical rehabilitation, assisted daily living (ADL), children and elderly care, has recently gained increasing attention. The emergence of some novel sensing devices featuring miniature size, a light weight, and wireless data transmission, the availability of wireless communication infrastructure, the progress of machine learning and deep learning algorithms, and the widespread IoT applications has promised new opportunities for a significant progress in this particular field. Motivated by a great demand for HAR-related applications and the lack of a timely report of the recent contributions to knowledge in this area, this investigation aims to provide a comprehensive survey and in-depth analysis of the recent advances in the diverse techniques and methods of human activity recognition and motion analysis. The focus of this investigation falls on the fundamental theories, the innovative applications with their underlying sensing techniques, data fusion and processing, and human activity classification methods. Based on the state-of-the-art, the technical challenges are identified, and future perspectives on the future rich, sensing, intelligent IoT world are given in order to provide a reference for the research and practices in the related fields.

Keywords: Internet of Things (IoT); human activity recognition (HAR); motion analysis; wearable sensors; machine learning

---

1. Introduction

In the future information systems featuring rich sensing, wireless inter-connection, and data analytics, which are usually denoted as Internet of Things (IoT) and cyber-physical systems (CPS) [1,2], human motions and activities can be recognized to perform intelligent actions and recommendations. The monitoring, recognition, and in-depth analysis of human gesture, posture, gait, motion, and other daily activities has received growing attention in a number of application domains, and significant progress has been made, attributing to the latest technology development and application demands [3]. As a result, new sensing techniques and mathematical methods are dedicated to the numerical analysis of the posture and activities of human body parts, and plenty of related studies are found in the literature.

The significant progress in human activity recognition (HAR) and motion analysis may be driven by the following powers: (1) The new powerful information systems require more efficient human–machine interactions for potential intelligence; (2) the development of integrated circuits (ICs) and circuits fabrication techniques has prompted the emergence of low-power, low-cost, miniature-sized, and flexible wearable sensing devices; (3) there have been recent advances in data analytics techniques, including filtering and machine learning techniques in dealing with data processing and classification; and (4) some new emerging applications fields, including wearable sensor-based medical care, sports analysis, physical rehabilitation, assisted daily living (ADL), etc., have also provided new opportunities for the techniques to revolutionize the traditional techniques in these fields. As a result, much research effort has been devoted to this area, and a number of technical solutions have been proposed.

There have been plenty of innovative studies that introduce novel sensing techniques to human activity recognition and motion analysis. Human–machine interaction (HMI) may be a critical area, where human gesture, posture, or motion can be recognized for the efficient interaction between machines covering a wide field of applications, including human–object interaction, virtual reality, immersive entertainment, etc. [4–6]. Human activity recognition and motion analysis has also been an effective way for sports analysis evaluation, while peer investigations are identified for golf swing analysis [7], swimming velocity estimation [8], and sports training [9]. Medical care has become a new research interest in the recent years, where precise and quantitative representation of human motion can help physicians in diagnosis, treatment planning, and progress evaluation; typical applications include gait analysis for stoke rehabilitation therapy [10], clinical finger movement analysis [11], Parkinson's disease treatment [12], etc. In addition, many other innovative applications are also found in assisted daily living, elderly and children care, etc. It is easy to see how through the applications, the innovative usage of the mentioned techniques has revolutionized the traditional approaches and resulted in convenience, efficiency, and intelligence that have never been seen before.

Many different techniques are employed to obtain the raw sensor data for monitoring human activities. The most commonly used techniques are the smartphone built-in inertial measurement units (IMUs) and optical cameras [13,14]. Since smartphones have almost become a must-have assistant for people's everyday life, a lot of human body and human activity related studies are carried out by taking advantage of smartphones. An optical camera, as a widely used sensing device, is a mainstream solution for human activity recognition. A depth camera, due to the one added dimension of depth information, has its unique strength compared to normal optical cameras [15]. Then, electromagnetic waves, such as frequency modulated continuous wave (FMCW) radar [16], impulse radio ultra-wide band (IR-UWB) radar [17], WiFi [18], and capacitive sensor array [19], featuring non-contact and non-line-of-sight (NLOS) traits, have also been introduced to human activity recognition. For the sensing techniques, accuracy, safety, privacy protection, and convenience are key factors for their applications. The low-cost, lightweight, and miniature-sized wearable IMU sensor devices have been prevalent techniques employed by peer investigations for human activity recognition, and the integration of two or more techniques may result in a better performance [20]. In addition, discrete sensing devices may be combined to constitute a wireless body area network (WBAN) for the establishment of wearable computing in many high-performance systems [21].

In addition to sensing techniques and networking techniques, data processing techniques including filtering, segmentation, feature extraction and classification are also indispensable enablers. For the pre-processing of sensing data, moving average filter (MAF), Kalman filter (KF) and complementary filter (CF) are the common approaches [22–24]. For classification, different machine learning algorithms have been used to create recognition models such as decision tree (DT), Bayesian network (BN), principle component analysis (PCA), support vector machine (SVM), artificial neural networks (ANNs), logistic regression, hidden Markov model (HMM), K-nearest neighbors (KNN) and deep neural networks (DNNs) [25–28]. Deep learning methods, such as the convolutional neural network (CNN) and recurrent neural network (RNN), due to their performance and wide acceptance in data analysis, have also recently gained interest in being used as tools [29–31]. The data processing techniques play an important role in guaranteeing the efficiency and accuracy of the recognition and analysis of human activities.

Admittedly, due to the potential in the different application fields, there has been a great demand for accurate human activity recognition techniques that can lead to the convenience, efficiency, and intelligence of information systems, new opportunities of medical treatment, more convenient daily living assistance, etc. However, there lacks a timely report on the recent contributions of recent technical advances with an in-depth analysis of the underlying technical challenges and future perspectives. Motivated by both the great demand for a more efficient interaction between human and information systems and the lack of investigations about the new contributions to knowledge in the field, this paper aims to provide a comprehensive survey and in-depth analysis of the recent advances in the diverse techniques and methods for human activity recognition and motion analysis.

The rest of this paper is organized as follows: Section 2 summarizes the innovative application regarding human activity recognition and motion analysis; Section 3 illustrates the fundamentals, including the common methodology, the modeling of human parts, and identifiable human activities. Sections 4 and 5 present the novel sensing techniques and mathematical methods, respectively. Then, Section 6 gives the underlying technical challenges and future perspectives, followed by Section 7, which concludes the work.

2. Latest Progress in HAR-Related Applications

The past few decades have witnessed unprecedented prosperity of electronics techniques and information systems, which have resulted in a revolutionary development in almost all aspects of technological domains, including aeronautics and astronautics, automotive industry, manufacturing and logistics, consumer electronics and entertainment, etc. Human activity recognition and motion analysis, due to its potential in wide areas of applications, has attracted much research interest and made remarkable progress in recent years. This section gives an overview of the latest technical progress and a summary of the application domains of human activity recognition and motion analysis.

2.1. Overview of the Latest Technical Progress

An overview of the latest technical progress of human activity recognition is given in Figure 1. The technical progresses of HAR are found to focus on the following aspects: new sensing device and methods, innovative mathematical methods, novel networking and computing paradigms, emerging consumer electronics, and convergence with different subject areas.

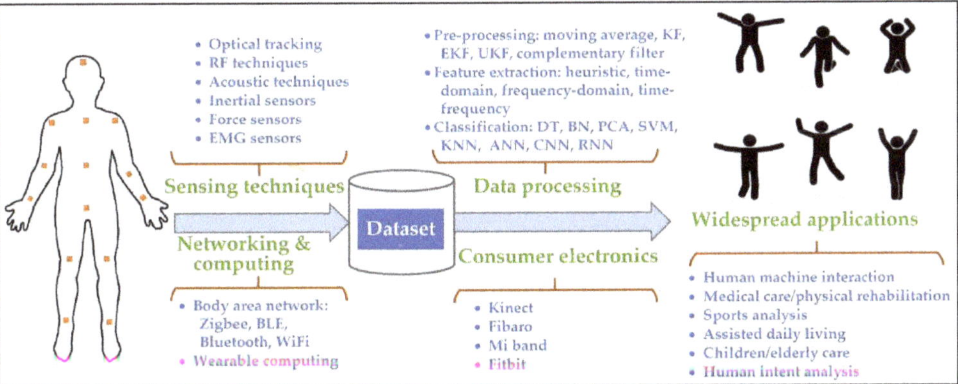

Figure 1. Overview of the latest technical progress of human activity recognition (HAR) investigations.

1. New sensing devices and methods: The acquisition of raw data is the first step for accurate and effective activity recognition. The cost reduction of the electronic devices has accelerated the pervasive sensing and computing systems, such as location and velocity tracking [32]. On account of the sensing techniques, many new techniques that were previously not possible for their cost,

size, and technical readiness are now introduced for human activity related studies in addition to the traditional video cameras (including depth cameras), FMCW radar, CW-doppler radar, WiFi, ultrasound, radio frequency identification (RFID), and wearable IMU and electromyography (EMG) sensors, etc. [15–18]. Among the above candidates, FMCW radar, CW-doppler radar, WiFi, ultrasound, and RFID are both NLOS and contactless. Wearable IMU is NLOS and body-worn, for which the applications are not limited to specific areas. The micro-electro-mechanical system (MEMS) IMUs, due to their advantages in being low-power, low-cost, miniature-sized, as well as able to output rich sensing information, have become a dominant technical approach for HAR studies [33,34].

2. Innovative of mathematical methods: To take advantage of the sensor data, an appropriate modeling of human body parts activities, the pre-processing of raw data, feature extraction, classification, and application-oriented analysis are pivotal enablers for the success of HAR-related functions. With respect to pre-processing, a number of lightweight, time-domain filtering algorithms, which are appropriate for resource-limited computing, including KF, extended Kalman filter (EKF), unscented Kalman filter (UKF), and Mahony complementary filter, are common alternatives that deal with wearable electronics [35,36]. In terms of feature extraction, the time-domain and frequency-domain features, including mean, deviation, zero crossing rate, and statistical characteristics of amplitude and frequency, are the most fundamental parameters [37]. Fast Fourier transform (FFT), as a classical algorithm, is the main solution for the frequency-domain feature analysis. With respect to classification, much effort in classical machine learning, including DT, BN, PCA, SVM, KNN, and ANN, and deep learning methods, including CNN and RNN, has been devoted to this particular area in recent years [25,38].
3. Novel networking and computing paradigms: The sensing and computing platforms are crucial factors for effective human activity recognition. For visual-based solutions, high-performance graphics processing units (GPUs) have paved the way for the intensive computations in HAR [39]. For wearable sensing solutions, many new pervasive and mobile networking and computing techniques for battery-powered electronics have been investigated. Various wireless networking and communication protocols, including Bluetooth, Bluetooth low energy (BLE), Zigbee, and WiFi were introduced for building a body area network (BAN) for HAR [40]. New computation paradigms, such as mobile computing and wearable computing, were proposed to handle the location-free and resource-limited computing scenarios, and sometimes the computation is conducted on an 8-bit or 16-bit micro-controller unit (MCU) [41]. The novel networking and computing paradigms customized for HAR are more efficient and flexible for the related investigation and practices.
4. Emerging consumer electronics for HAR: Another evidence of the progress in HAR is the usage of emerging HAR consumer electronics, including Kinect, Fibaro, the Mi band, and Fitbit. The Kinect-based somatosensory game is a typical use case of recognizing human motion as an input for human–computer interactions [42]. Fibaro detects human motion and changes in location for smart home applications [43]. Some wearable consumer electronics, such as the Mi band and Fitbit, can provide users with health and motion parameters, including heartbeats, intensity of exercises, walking or running steps, sleeping quality evaluation, etc. [44,45]. In addition, there are electronic devices such as MAX25205 (Maxim, San Jose, USA) used for gesture sensing in automotive applications using an IR-based sensor, where hand swipe gestures, finger, and hand rotation can be recognized [46]. These devices have stepped into people's daily lives to assist people to better understand their health and physical activities, or to perform automatic control and intelligent recommendation via HAR.
5. Convergence with different subject areas: The HAR techniques were also found to be converging with many other subject areas and thus continually allowing for new applications to be created. Typically, HAR is merging with medical care, and this has resulted in some medical treatment and physical rehabilitation methods of diseases such as stroke and Parkinson's disease [10,12].

HAR has also been introduced in sports for sports analysis for the purpose of enhancing athletes' performance [7,8]. HAR-assisted daily living is another example as a field of application, where power consumption, home appliance control, and intelligent recommendations can be implemented to customize the living environment to people's preferences [2,47].

2.2. Widespread Application Domains

Driven by the subtantial technical progress in the related fields, HAR has been extended to a wide spectrum of application domains. Since human involvement is the most critical part for many information systems, the introduction of HAR can potentially result in greater efficiency and intelligence of the interaction between human and information systems, which also creates new opportunities for the human body or human activity related studies and applications. The fields of applications/domains are summrized as follows:

1. Human–machine interaction (HMI)—HAR is used to recognize the gesture, posture, or motion of human body parts and use the results as inputs for information systems, and to perform efficient interactions or provide people intelligent recommendations [48,49].
2. Medical care/physical rehabilitation—HAR is used to record the motion trajectory of human body parts, conduct numerical analysis, and the results can be used as reference for medical treatment [12,50].
3. Sports analysis—HAR is used to record information such as the acceleration, angular velocity, velocity, etc., for numerical analysis of human body parts motions in order to evaluate the performance of sports and advise improvement strategies [7,51].
4. Assisted daily living—HAR is used to recognize people's gesture and motions for home appliance control, or analyze people's activities and habits for living environment customizations and optimization.
5. Children and elderly care—HAR is used to analyze children's or elderly people's activities, such as fall detection and disease recognition and to perform assistance or report warning to caregivers [52,53].
6. Human intent analysis—HAR is used to analyze people's motions and activities in order to predict people's intents in a passive way for particular applications, such as intelligent recommendation, crime detection in public areas, etc. [54].

Although many new techniques and methods have been introduced in HAR for different applications and remarkable progress has been made, there is still large room for improving the performance of the methods and systems. More convenient, lightweight, powerful computation sensing devices and systems, more accurate classification algorithms, and some application-oriented studies will still continually gain attention and play a more important role in various information systems and in people's daily lives.

3. Fundamentals

In order to give a comprehensive understanding of the HAR techniques and methods, this section presents the fundamentals of HAR regarding the sensing devices and data processing techniques. Since there are many different sensing devices, pre-processing, and classification techniques involved, the abstract, common methodology covering the different techniques and methods is illustrated, and the correspondence between the identifiable human activities with the innovative applications is reported as well.

3.1. Common Methodology

In view of the sensing and processing techniques for HAR, the methods can be described with a common methodology as shown in Figure 2, which may be applicable for both optical tracking and wearable sensing techniques [55,56]. The common methodology consists of four steps: data acquisition and pre-processing, segmentation, feature extraction, and classification [57].

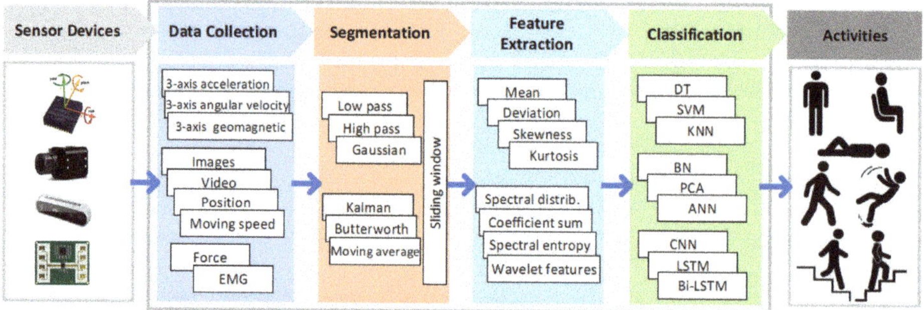

Figure 2. Common methodology for HAR.

1. Data acquisition and pre-processing—Motion sensors such as 3D cameras, IMU, IR-radar, CW radar, and ultrasound arrays, either mounted onto a fixed location or body-worn, are used to obtain raw data including human activity information, such as position, acceleration, velocity, angular velocity, and angles. In addition, various noise cancellation techniques, either time-domain or frequency-domain ones, need to be applied to optimize the quality of the signal for further processing.
2. Segmentation—Human motion is normally distributed in particular time spans, either occasionally or periodically. The signals need to be split into window segments for activity analysis, and the sliding window technique is usually used to handle the activities between segments.
3. Feature extraction—The features of activities including joint angle, acceleration, velocity and angular velocity, relative angle and position, etc., can be directly obtained for further analysis. There may also be indirect features that can be used for the analysis of particular activities.
4. Classification/model application—The extracted features can be used by various machine learning and deep learning algorithms for the classification and recognition of human activities. Some explicit models, such as skeleton segment kinematic models, can be applied directly for decision-making.

3.2. Modeling of Human Body Parts for HAR

Since the human body is a systematic whole, a primary step in using HAR is to build a mechanical and posture kinematic model for the motion of human body parts. Human skeleton joints are the key features for optical tracking based HAR methods, where markers are sometimes applied for human gesture and posture recognition as well as motion tracking [53,58]. Figure 3 shows a skeleton-based multi-section model for the hand and human body. For wearable IMU-based HAR, the multi-segment model of human skeleton is also used as a basis to build a kinematic model of body segments. For example, the kinematic model of an elbow joint can be estabished with two segments connected by two revolute joints, allowing two degrees of freedom (DoF), as shown in Figure 4 [59]; the DoF kinematic model of the left leg, including three DoF angle joints, one DoF knee joint, and three DoF hip joints, is presented in [60], and the kinematics of the ankle to the hip and trunk is discussed in [61]. With the multi-segment model and kinematic model, the posture and the motion of human body parts and the whole body can be estimated with the angle and position data obtained with sensor devices.

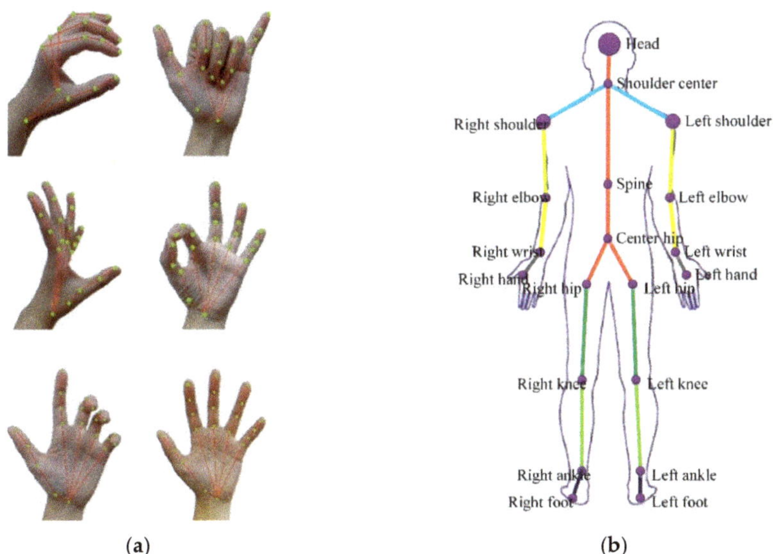

Figure 3. Multi-segments model for human body: (**a**) hand skeletal model, (**b**) whole body skeletal model.

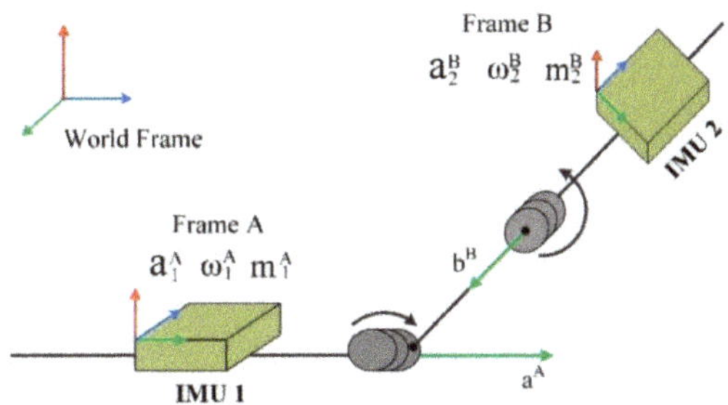

Figure 4. Kinematic model of elbow joint.

3.3. *Identifiable Human Activities*

Since human body parts are not rigid structures and are, instead, structures that deform and move in different ways following complex kinematic patterns, human activity parameters that belong to these parts are very hard to measure. Many new technical solutions are customized for human activity measurment, and remarkable progress has been made. The identifiable human activities are summarized in Table 1.

Table 1. Identifiable human activities and example applications.

Human Body Parts	Activities	Example Applications
Hand	Finger movement	Rheumatoid arthritis treatment [11]
	Hand gesture	Robot-assisted living [62]
	Hand movement	Handwriting recognition [63]
Upper limb	Forearm movement	Home health care [64]
	Arm swing	Golf swing analysis [7]
	Elbow angle	Stroke rehabilitation [59]

Table 1. *Cont.*

Human Body Parts	Activities	Example Applications
Lower limb	Gait analysis	Gait rehabilitation training [10]
	Knee angles in movement	Risk assessment of knee injury [60]
	Ankle movement	Hip and trunk kinematics [61]
	Stairs ascent/descent	Human activity classification [65]
Spine	Swimming	Swimming motion evaluation [51]
Whole body	Fall detection	Elderly care [18]
	Whole-body posture	Bicycle riding analysis [66]
	Whole-body movement	Differentiating people [67]

As shown in Table 1, many human body parts and the human body as a whole can be used for activity recognition for different purposes, including fingers, hand, arm, elbow, knee, ankle, and the whole body. Of course, the identifiable human activities and the corresponding applications are not limited to those listed in the table. Rapid technical progress has resulted in the continual addition of new possibilities and the creation of new applications in this particular field.

4. Novel Sensing Techniques

The appropriate measurement techniques and quality sensor data are the fundamentals for effective HAR and further applications. In this section, the sensing techniques for HAR are summarized with taxonomy and compared with proposed indexes. The sensor network solutions for wireless data collection and multi-sensor nodes for whole body monitoring are also discussed.

4.1. Taxonomy and Evaluation

According to the principles of sensor devices, the sensing techniques can be divided into six categories: optical tracking, radio frequency techniques, acoustic techniques, inertial measurement, force sensors, and EMG sensors. Each of the above categories may include a couple of different sensing techniques that may exhibit different performances in dealing with HAR applications. The feasibility of sensing techniques for HAR can be evaluated with five indicators: convenience of use, application scenarios, richness of information, the quality of raw data, and cost. The discussion and analysis of the sensing techniques are conducted with the abovementioned categories and evaluation indicators, which are as shown in Figure 5.

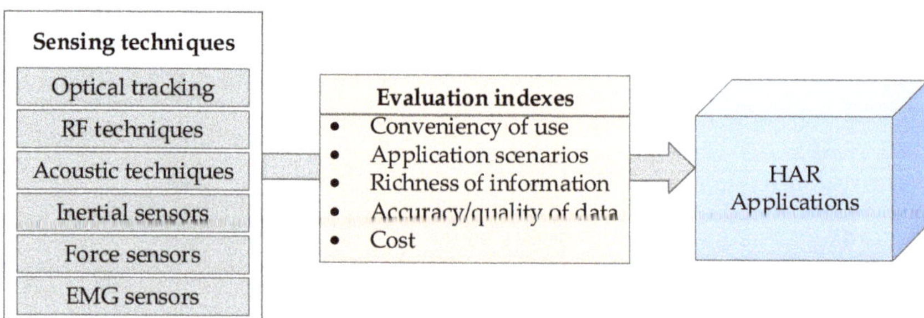

Figure 5. Taxonomy and evaluation indexes for HAR sensing techniques.

4.2. Sensing Techniques with Example Applications

To give a comprehensive overview and in-depth discussion of the sensing techniques for HAR and related studies, the sensing techniques of the five types and the corresponding typical applications are summarized in Table 2.

Table 2. Sensing techniques with typical applications.

Type	Sensing Techniques	Typical Applications
Optical	RGB camera	Finger spelling recognition [68], detection of hand trajectories for industrial surveillance [69], gait measurement at home [70]
	RGB-D	Hand gesture recognition [51], human behavior understanding [71], human body reshaping [72]
	Depth sensor	3D dynamic gesture recognition [58], real-time continuous action recognition [73], full hand pose recognition [15]
RF	IR-UWB [1]	Non-contact human gait identification [17], non-contact human motion recognition [74], human respiration pattern recognition [75]
	FMCW radar	Continuous human activity classification [16,76], human limb motion detection [77]
	Commodity WiFi	Contactless fall detection [18], handwritting recognition [63,78], gesture and user detection [79]
	UHF [2] RFID	Ambient assisted living [80], in-car activity recognition [81]
Acoustic	Ultrasound	Lower limb motion estimation [82], hand gesture recognition [83], finger motion perception and recognition [84]
Inertial	3-axis accelerometer	Discrimination of human activity [85], arm movement classification for stroke rehabilitation [64]
	6- or 9-axis IMU	Recognition of daily and sport activity [86], hip and trunk motion estimation [61], assessment of leg squat [60], measurement of elbow angles for physical rehabilitation [59], reconstruction of angular kinematics for smart home healthcare [87], fall detection and daily activity recognition [88]
Force	Piezoelectric, FSR [3], thin film strain sensor, etc.	Wearable gait recognition [89], human motion detection [90], real-time gait feedback and training [91]
EMG	EMG electrodes	Classification of finger, hand, wrist, and elbow gestures [92]; knee movement estimation [93]; wrist-hand pose identification [94]

[1] IR-UWB, [2] UHF, and [3] FSR are the abbreviations of impulse radio ultra-wide band, ultra-high frequency, and force sensitive resistor, respectively.

1. Optical tracking—Optical tracking with one or a few common cameras is the most traditional way of human activity recognition. The temporal and spatial features of the human body and the activities can be extracted with image processing algorithms. The classification that takes advantage of the features can be sorted out with machine learning or deep learning techniques. This method is competitive in accuracy and wide application domains, including entertainment, industrial surveillances, public security, etc. With the progress in high-speed cameras and depth sensors in recent years, more innovative investigations and applications are found in both academic studies and practical applications. The strengths of optical tracking methods are high accuracy, contactless monitoring, and rich activity information, while the weaknesses are the fixed location of use and potential risk of privacy leak.
2. RF and acoustic techniques—The sensing technique working under the principle of radar includes IR-UWB, FMCW radar, WiFi, and ultrasound. The motion of human body parts or the whole body can be perceived using Doppler frequency shift (DFS), and some time–frequency signatures can be obtained for further analysis. Machine learning and deep learning techniques are widely used for classifications of human activities. Evidently, the strengths of this method are contactless measurement, NLOS, and no risk of privacy leak. Specifically, the commodity WiFi takes advantage of the existing communication infrastructure without added cost. Their

weaknesses are the monotonousness and implicity of information provided by the sensors which requires specialized processing.

3. Inertial sensors—The inertial sensors, especially the MEMS IMUs, became a dominant technical approach for human activity recognition and analysis. They provide 3-axis acceleration, 3-axis angular velocity, and 3-axis magnetometer signals, which can be employed for the estimation of attitude and motion of human body parts by mounting the devices on them. The strengths of IMUs in HAR are their miniature size, low cost, low power, and rich information output, which make them competitive for wearable applications that can be used without location constraints. Their weakness is their contact measurement, which may be inconvenient to people for daily activities recognition.
4. Force sensors and EMG sensors—Force sensors may include piezoelectric sensors, FSR, and some thin film pressure sensors of different materials. They may provide pressure or multiple axis forces of human gait, hand grasp, etc., for sports analysis, physical rehabilitation or bio-feedback. EMG sensors work in a similar way as force sensors do and implement similar functions by using the EMG signal. Since EMG provides the muscle activities, it may provide more useful information for medical care, such as rehabilitation and amputation. The strengths of force sensors and EMG sensors are their capability in obtaining useful information of body part local areas. Their weakness is also the contact measurement, which may be inconvenient to people for daily activities recognition.
5. Multiple sensor fusion—In addition to the above, there are also multiple sensor fusion based techniques that combine more than one of the above alternatives. For instance, inertial and magnetic sensors are combined for body segments motion observation [24]; optical sensor, IMU, and force sensor are combined for pose estimation in human bicycle riding [66]; knee electro-goniometer, foot-switches, and EMG are combined for gait phase and events recognition [95]; and FMCW radar and wearable IMUs are combined for fall detection [96]. The purpose of the combination is to compensate the limitations of one sensor technology by taking advantage of another so as to pursue maximal performance. For example, IMU and UWB radar are usually combined with real-time data fusion to overcome the limitations in continuous error accumulation of IMU, and NLOS shadowing and random noise of UWB [23].

On account of the different sensing techniques, they may behave differently with respect to the different evaluation indexes. In summary, for the convenience of use, the optical tracking and radar techniques are non-contact, but they are normally installed on particular locations. The inertial sensors can be wearable and provide rich useful information for human activity analysis, but contact measurement is involved, which may introduce inconvenience for users' daily activities. As a contact measurement method, force sensors and EMG can reveal information of body part local areas, which is competitive in medical care.

4.3. Body Area Sensor Networks

For the wearable sensing techniques, it is important to carry out the data collection without interrupting people's normal activities. Therefore, power cords data wires should be replaced with miniature batteries and wireless communication nodes. The optional wireless communication techniques which are commonly used include Bluetooth, WiFi, BLE, Zigbee, 4G/5G, NB-IoT, and LoRa, etc. [97] Since 4G/5G, NB-IoT, and LoRa are commonly used for long range data transmission for different purposes, Bluetooth, WiFi, BLE, and Zigbee are more likely to be chosen for short range wearable communications. A comparison of the wireless techniques is given in Table 3.

Since wearable sensing devices are powered with batteries, the energy consumption, data rate, and network topology are the key parameters to be considered when establishing the systems. Normally, Bluetooth is used for the data collection of single sensor devices: For example, Bluetooth is employed to transmit IMU data for sports analysis [7], force data of piezoelectric sensors for gait recognition [89], and IMU data for assessment of elbow spasticity [98]. Then, BLE and ZigBee are competitive in building a BAN for the data collection of multiple nodes systems; ZigBee is used for multiple IMU data collection in [66], while BLE-based wireless BAN (WBAN) was used for joint angle estimation in [99].

Table 3. Alternative wireless communication standards for HAR.

Standards	Range	Maximum Data Rate	Carrier Frequency	Energy Consumption	Network Topology
Bluetooth (Ver. 4.2)	10 m	2 Mbps	2.4 GHz	Medium	P2P
BLE	10 m	1 Mbps	2.4 GHz	Low	Star, mesh
ZigBee	50 m	250 Kbps	2.4 GHz	Low	Star, tree, mesh
WiFi (802.11 n)	10–50 m	450 Mbps	2.4 GHz	High	Star, tree, mesh, P2P

5. Mathematical Methods

Once the sensor data are obtained, how to make use of it is also a challenging task. As presented in the common methodology given in Figure 2, the key steps include pre-processing, segmentation, feature extraction, and classification/model application. In this section, we focus on the mathematical methods for the two key steps: feature extraction and classification.

5.1. Features and Feature Extraction

The feature extraction methods can be categorized into four types according to their outputs: heuristic, time-domain, time–frequency, or frequency-domain features [85,100], which are as shown in Table 4. Heuristic features are those derived and characterized by an intuitive understanding of how activities produce signal changes. Time-domain features are typically statistical measures obtained from a windowed signal that are not directly related to specific aspects of individual movements or postures. Time–frequency features, such as wavelet coefficients, are competitive for detecting the transitions between human activities [65]. Then, frequency-domain features are usually the preferred option for human activity recognition. Mathematical tools such as FFT, short-time Fourier transform (STFT), and discrete cosine transform (DCT) are the commonly used methods.

Table 4. Types of features for HAR [60,71,80,95].

Types	Examples of Features
Heuristic	Minimum, maximum, zero-crossing, threshold amplitude, inclination angles, signal magnitude area (SMA), etc.
Time domain	Mean, variance, standard deviation (SD), skewness, kurtosis, root mean square (RMS), median absolute square (MAS), interquartile range, minimum, interquartile range, auto-correlation, cross correlation, motion length, motion cycle duration, kinematic asymmetry, etc.
Frequency domain	Spectral distribution, coefficients sum, spectral entropy, etc.
Time–frequency	Approximation coefficients, detail coefficients, transition times

5.2. Classification and Decision-Making

For most human activity recognition studies, the classification to discriminate the different types of activities using the extracted features is a critical step. Due to the complexity in human gesture, posture, and daily activities, how to descriminiate them with certain accuracy using different mathematical models becomes a research focus. The classification can be classified into two categories:

1. Threshold-based method—Threshold can be eaisly used to obtain many simple gestures, postures, and motions. For example, fall detection, hand shaking, and static standing can be recognized using the acceleration threshold calculated using equation $a_{TH} = \sqrt{a_x^2 + a_y^2 + a_z^2}$, and walking or running can be recognized using Doppler frquency shift thresholds.

2. Machine learning techniques—Machine learning is a method of data analysis that automates analytical model building, which is suitable for the classification of human activities using sensor data. Its feasibility and efficiency have been demonstrated by many published studies with accuracies over 95%. Typically, HMM achieves an accuracy of 95.71% for gesture recognition [101]; SVM and KNN are employed for human gait classification between walking and running with an accurate over 99% [102]; PCA and SVM recognize three different actions for sports training with an accuracy of 97% [9]; KNN achieves an accuracy of 95% for dynamic activity classification [65]. There are also peer investigations providing evaluations of different machine learning methods [9,88,103].
3. Deep learning methods—Deep learning techniques take advantages of many layers of non-linear information processing for supervised or unsupervised feature extraction and transformation, as well as for pattern analysis and classification. They presents advantages over traditional approaches and have gained continual research interest in many different fields, including HAR in recent years. For example, CNN achieves an accuracy of 97% for the recognition of nine different activities in [31], an accuracy of 95.87% for arm movement classification [64], an accuracy of 99.7% for sEMG-based gesture recognition in [93], and an accuracy of 95% for RFID based in-car activity recognition in [81]. The past investigations have demonstrated that deep learning technique is an effective way of classification for human activity recognition. There are also investigations applying long short-term memory (LSTM) [104] or bidirectional LSTM (Bi-LSTM) [76,96], as well as recurrent neural networks in sports and motion recognition, which have exhibited competitive performance compared to CNN.

Due to the superior performance of machine learning and deep learning techniques, many tool libraries—such as MATLAB machine learning toolbox, pandas, and scikit-learn [105]—for different development environments are available, which makes the implementation of the classification quite convenient. The different machine learning techniques will continually be considered the dominant tool for classification for various sensor data based human activity recognition.

6. Discussions

Driven by the related sensing, communication, and data processing techniques, HAR has undergone a rapid progress and is extended to widespread fields of applications. This section summarizes the state-of-the-art, the underlying technical challenges, and the future trend.

6.1. Performance of the State-of-the-Art

Based on the above investigation, it is a critical issue to determine how to select the appropriate sensing techniques and mathematical methods for HAR applications. It is found that the technical solutions are highly dependent on the particular HAR tasks. For accurate indoor human activity recognition, optical depth sensors with PCA, HMM, SVM, KNN, and CNN are commonly seen to obtain an accuracy over 95%, and combination of UWB and IMU can also find applications with accuracy over 90%. For wearable outdoor applications of consumer electronics, sports analysis, and physical rehabilitation, a single IMU is a competitive choice, which normally uses Kalman filter or a complementary filter for noise cancellation and uses SVM or CNN for decision-making. The accuracy for walking and running step counting and other motion recognition is normally over 90%. Then, for medical care, physical rehabilitation or bio-feedback, thin film force sensors and EMG sensors are commonly used to obtain information of local areas of body parts. Deep learning is the most prevalent choices, which may result in an accuracy at about 90%. According to the analysis in Sections 4 and 5, each sensing technique and mathematical method has its own pros and cons and suitable application scenarios. The combination of two or more sensing techniques or mathematical methods may overcome the limitations of one technique by taking advantage of the others.

6.2. Technical Challenges

According the survey of the state-of-the-art, the techniques for HAR are far from technically ready for various application in the different application fields. Eventually, the technical challenges for the human activity recognition and motion analysis techniques are summarized as follows:

1. Limitation of sensing techniques for convenient uses—The sensing techniques, including optical tracking, radar, inertial measurement, force, and EMG sensors, all have their limitations in performing accurate, non-contact, location-free human activity sensing. Optical sensors are competitive in accuracy, but they are LOS and limited to specific areas. Radar techniques, such as FMCW, IR-UWB, WiFi, and ultrasound, are NLOS, but they work in specific areas, and they are not as accurate. IMUs, force data, and EMG are low cost and low power, making them suitable for wearable applications, but the battery power and wireless modules make the size of the devices inconvenient for daily use. They lack sensing techniques that do not interrupt people's normal daily activities.
2. Dependency on PCs for data processing—For the body-worn sensing devices, the data acquisition and pre-processing are usually completed with low power MCUs, and further processing such as feature extraction and classification are conducted on PCs. This results in the requirements of a high transmission data rate and the involvement of a powerful PC. The high data rate is also a challenge for the deployment of multiple sensor nodes to establish a BAN. The traditional way of communication and computation has constrained the application of body-worn HAR.
3. Difficulties in minor gesture recognition—Most of the existing techniques are for the recognition of regular and large-scale human gesture, posture, and movements, such as hand gesture, arm swing, gait, and whole-body posture, with classification techniques. It is still a challenge to quantitively discriminate minor gesture variations, which may be potentially valuable for many applications.
4. Specificity in human activity recognition and motion analysis—Most of the state-of-the-art may focus on the postures or activities for a particular body part, or a particular posture or activity of the whole body. The focus usually falls on the specific sensing devices and classification techniques, while lacks comprehensive posture and activity recognition and further understanding of the kinematics of human body and behaviors of people based on the in-depth analysis of their daily activities.

6.3. Future Perspective

To overcome the technical challenges mentioned above, a number of future investigations are expected to focus on the following aspects:

1. Unobtrusive human activity monitoring techniques—The sensing techniques that can perform an unobtrusive perception of human activities without or with minimum interferece to people's normal activities, and that can also be used anywhere, need greater study. New investigations about minimizing the size of sensing devices and the way of powering the device may be new research interests.
2. Optimization of sensing network and computation powers—It is critical to explore the optimal BAN of sensing devices for HAR, and to optimize the computation power between sensor nodes and the central processor to relieve the burden of the sensor network for data transmission. A new paradigm of wearable computing for wearable BAN may become an effort attracting topic.
3. Further studies for minor gestures and posture recognition—Further investigation of the sensing techniques and feature extraction techniques for minor gesture recognitions may find applications in different fields, such as HMI. For example, UWB Doppler features based human finger motion recognition for the operation of handheld electronics may be one potential use case.
4. Comprehensive recognition and understanding of human activities—Based on the recognition of human gestures and movements, the further positional and angular kinematics of the human

body and human behaviors can be investigated for the further analysis. Research outputs may provide valuable reference for the studies of bionic robotics and for customized human intent estimation as well as smart recommendations in smart environments.

The above technical challenges may draw attention to the investigations of this field in the future, and the related studies about unobtrusive sensing, optimized sensing network and computing system, minor gesture and posture recognition, and further human body kinematics and behavior understanding are promising research topics that may add new features to HAR and create new opportunities of research and applications. Since HMI is a key attribute for the usability of information systems, HAR as a promising solution for the efficient and convenient interaction between human and inforamtion systems will play an important role in the future rich, sensing IoT world.

7. Conclusions

This article presented a comprehensive overview of the recent progress in the sensing techniques that have been introduced to human activity recognition and motion analysis. Remarkable technical progresses in new sensing devices and methods, innovative mathematical models for feature extraction and classification, novel networking and computing paradigms, convergence with different subject areas have been identified, which have extended to widespread application fields. To provide a comprehensive understanding of the fundamentals, the skeletal based multi-segment models and kinematic modeling of human body parts were firstly presented. Then, the sensing techniques were summarized and classified into six categories: optical tracking, RF sensors, acoustic sensors, inertial sensors, force sensors, and EMG sensors, followed by in-depth discussions about the pros and cons of the proposed evaluation indexes. Further to the sensing devices, the mathematical methods including feature extraction and classification techniques are discussed as well. According to the state-of-the-art HAR techniques, the technical challenges focused on were found to include the limitation of sensing techniques for convenient uses, dependency on PCs for data processing, difficulties in minor gesture recognition, and specificity in human activity recognition and motion analysis. The solutions for these challenges are considered to be the development trend of future studies. Since human activity recognition and moton analysis is a promising way for efficient interaction between human and information systems, it may play a more important role in future IoT enabled intelligient information systems.

Author Contributions: Conceptualization, Z.M. and M.Z.; methodology, M.Z. and H.Z.; formal analysis, Z.M.; investigation, M.Z., C.G., and Q.F.; resources, Z.M.; writing—original draft preparation, M.Z., C.G., and Q.F.; writing—review and editing, Z.M. and H.Z.; visualization, supervision, N.G. and Z.Z.; project administration, Z.M.; funding acquisition, Z.M. All authors have read and agreed to the published version of the manuscript.

Funding: The work presented in this paper is supported by the National Natural Science Foundation of China (NSFC) (51805143), Natural Science Foundation of Hebei province (E2019202131), and the Department of Human Resources and Social Security of Hebei Province (E2019050014 and C20190324).

Acknowledgments: The authors would like to thank Nondestructive Detection and Monitoring Technology for High Speed Transportation Facilities, Key Laboratory of Ministry of Industry, and Information Technology for support.

Conflicts of Interest: The authors declare no conflict of interest.

References

1. Dian, F.J.; Vahidnia, R.; Rahmati, A. Wearables and the Internet of Things (IoT), Applications, Opportunities, and Challenges: A Survey. *IEEE Access* **2020**, *8*, 69200–69211. [CrossRef]
2. Aksanli, B.; Rosing, T.S. Human Behavior Aware Energy Management in Residential Cyber-Physical Systems. *IEEE Trans. Emerg. Top. Comput.* **2017**, 845–857. [CrossRef]
3. Chen, L.; Hoey, J.; Nugent, C.D.; Cook, D.J.; Yu, Z. Sensor-based Activity Recognition. *IEEE Trans. Syst. ManCybern.* Part C Appl. Rev. **2012**, *42*, 790–808. [CrossRef]

4. Ziaeefard, M.; Bergevin, R. Semantic Human Activity Recognition: A Literature Review. *Pattern Recognit.* **2015**, *48*, 2329–2345. [CrossRef]
5. Zhang, F. Human–Computer Interactive Gesture Feature Capture and Recognition in Virtual Reality. *Ergon. Des. Q. Hum. Factors Appl.* **2020**. [CrossRef]
6. Vrigkas, M.; Nikou, C.; Kakadiaris, I.A. A review of Human Activity Recognition Methods. *Front. Robot. AI* **2015**, *2*, 28. [CrossRef]
7. Kim, Y.J.; Kim, K.D.; Kim, S.H.; Lee, S.; Lee, H.S. Golf Swing Analysis System with a Dual Band and Motion Analysis Algorithm. *IEEE Trans. Consum. Electron.* **2017**, *63*, 309–316. [CrossRef]
8. Dadashi, F.; Millet, G.P.; Aminian, K. Gaussian Process Framework for Pervasive Estimation of Swimming Velocity with Body-worn IMU. *Electron. Lett.* **2013**, *49*, 44–46. [CrossRef]
9. Wang, Y.; Chen, M.; Wang, X.; Chan, R.H.M.; Li, W.J. IoT for Next Generation Racket Sports Training. *IEEE Internet Things J.* **2018**, *5*, 4559–4566. [CrossRef]
10. Wang, L.; Sun, Y.; Li, Q.; Liu, T.; Yi, J. Two Shank-Mounted IMUs-based Gait Analysis and Classification for Neurological Disease Patients. *IEEE Robot. Autom. Lett.* **2020**, *5*, 1970–1976. [CrossRef]
11. Connolly, J.; Condell, J.; O'Flynn, B.; Sanchez, J.T.; Gardiner, P. IMU Sensor-Based Electronic Goniometric Glove for Clinical Finger Movement Analysis. *IEEE Sens. J.* **2018**, *18*, 1273–1281. [CrossRef]
12. Nguyen, H.; Lebel, K.; Bogard, S.; Goubault, E.; Boissy, P.; Duval, C. Using Inertial Sensors to Automatically Detect and Segment Activities of Daily Living in People with Parkinson's Disease. *IEEE Trans. Neural Syst. Rehabil. Eng.* **2018**, *26*, 197–204. [CrossRef] [PubMed]
13. Mukhopadhyay, S.C. Wearable Sensors for Human Activity Monitoring: A Review. *IEEE Sens. J.* **2015**, *15*, 1321–1330. [CrossRef]
14. Lara, O.D.; Labrador, M.A. A Survey on Human Activity Recognition Using Wearable Sensors. *IEEE Commun. Surv. Tutor.* **2013**, *15*, 1192–1209. [CrossRef]
15. Rossol, N.; Cheng, I.; Basu, A. A Multisensor Technique for Gesture Recognition Through Intelligent Skeletal Pose Analysis. *IEEE Trans. Hum. Mach. Syst.* **2016**, *46*, 350–359. [CrossRef]
16. Vaishnav, P.; Santra, A. Continuous Human Activity Classification with Unscented Kalman Filter Tracking Using FMCW Radar. *IEEE Sens. J.* **2020**, *4*, 7001704. [CrossRef]
17. Rana, S.P.; Dey, M.; Ghavami, M.; Dudley, S. Non-Contact Human Gait Identification Through IR-UWB Edge-based Monitoring Sensor. *IEEE Sens. J.* **2019**, *19*, 9282–9293. [CrossRef]
18. Wang, H.; Zhang, D.; Wang, Y.; Ma, Y.; Li, S. RT-Fall: A Real-Time a Contactless Fall Detection Systems with Commodity WiFi Devices. *IEEE Trans. Mob. Comput.* **2017**, *16*, 511–526. [CrossRef]
19. Tariq, O.B.; Lazarescu, M.T.; Lavagno, L. Neural Networks for Indoor Human Activity Reconstructions. *IEEE Sens. J.* **2020**. [CrossRef]
20. Li, T.; Fong, S.; Wong, K.K.L.; Wu, Y.; Yang, X.-S.; Li, X. Fusing Wearable and Remote Sensing Data Streams by Fast Incremental Learning with Swarm Decision Table for Human Activity Recognition. *Inf. Fusion* **2020**, *60*, 41–64. [CrossRef]
21. Paoletti, M.; Belli, A.; Palma, L.; Vallasciani, M.; Pierleoni, P. A Wireless Body Sensor Network for Clinical Assessment of the Flexion-Relaxation Phenomenon. *Electronics* **2020**, *9*, 1044. [CrossRef]
22. Baldi, T.; Farina, F.; Garulli, A.; Giannitrapani, A.; Prattichizzo, D. Upper Body Pose Estimation Using Wearable Inertial Sensors and Multiplicative Kalman Filter. *IEEE Sens. J.* **2020**, *20*, 492–500. [CrossRef]
23. Zhang, H.; Zhang, Z.; Gao, N.; Xiao, Y.; Meng, Z.; Li, Z. Cost-Effective Wearable Indoor Localization and Motion Analysis via the Integration of UWB and IMU. *Sensors* **2020**, *20*, 344. [CrossRef] [PubMed]
24. Fourati, H.; Manamanni, N.; Afilal, L.; Handrich, Y. Complementary Observer for Body Segments Motion Capturing by Inertial and Magnetic Sensor. *IEEE ASME Trans. Mechatron.* **2014**, *19*, 149–157. [CrossRef]
25. Ferrari, A.; Micucci, D.; Mobilio, M.; Napoletano, P. On the Personalization of Classification Models for Human Activity Recognition. *IEEE Access* **2020**, *8*, 32066. [CrossRef]
26. Stikic, M.; Larlus, D.; Ebert, S.; Schiele, B. Weakly Supervised Recognition of Daily Life Activities with Wearable Sensors. *IEEE Trans. Pattern Anal. Mach. Intell.* **2011**, *33*, 2521–2537. [CrossRef]
27. Ghazal, S.; Khan, U.S.; Saleem, M.M.; Rashid, N.; Iqbal, J. Human Activity Recognition Using 2D Skeleton Data and Supervised Machine Learning. *IET Image Process.* **2019**, *13*, 2572–2578. [CrossRef]
28. Manzi, A.; Dario, P.; Cavallo, F. A human Activity Recognition System based on Dynamic Clustering of Skeleton Data. *Sensors* **2017**, *17*, 1100. [CrossRef]

29. Plotz, T.; Guan, Y. Deep Learning for Human Activity Recognition in Mobile Computing. *Computer* **2018**, *51*, 50–59. [CrossRef]
30. Xu, C.; Chai, D.; He, J.; Zhang, X.; Duan, S. InnoHAR: A Deep Neural Network for Complex Human Activity Recognition. *IEEE Access* **2019**, *7*, 9893. [CrossRef]
31. Bianchi, V.; Bassoli, M.; Lombardo, G.; Fornacciari, P.; Mordonini, M.; Munari, I.D. IoT Wearable Sensor and Deep Learning: An Integrated Approach for Personalized Human Activity Recognition in a Smart Home Environment. *IEEE Internet Things J.* **2019**, *6*, 8553–8562. [CrossRef]
32. Yuan, Q.; Chen, I.-M. Localization and Velocity Tracking of Human via 3 IMU Sensors. *Sens. Actuators A Phys.* **2014**, *212*, 25–33. [CrossRef]
33. Tian, Y.; Meng, X.; Tao, D.; Liu, D.; Feng, C. Upper Limb Motion Tracking with the Integration of IMU and Kinect. *Neurocomputing* **2015**, *159*, 207–218. [CrossRef]
34. Zihajehzadeh, S.; Loh, D.; Lee, T.J.; Hoskinson, R.; Park, E.J. A Cascaded Kalman Filter-based GPS/MEMS-IMU Integrated for Sports Applications. *Measurement* **2018**, *73*, 200–210. [CrossRef]
35. Tong, X.; Li, Z.; Han, G.; Liu, N.; Su, Y.; Ning, J.; Yang, F. Adaptive EKF based on HMM Recognizer for Attitude Estimation Using MEMS MARG Sensors. *IEEE Sens. J.* **2018**, *18*, 3299–3310. [CrossRef]
36. Enayati, N.; Momi, E.D.; Ferrigno, G. A Quaternion-Based Unscented Kalman Filter for Robust Optical/Inertial Motion Tracking in Computer-Assisted Surgery. *IEEE Trans. Instrum. Meas.* **2015**, *64*, 2291–2301. [CrossRef]
37. Wagstaff, B.; Peretroukhin, V.; Kelly, J. Robust Data-Driven Zero-Velocity Detection for Foot-Mounted Inertial Navigation. *IEEE Sens. J.* **2020**, *20*, 957–967. [CrossRef]
38. Jia, H.; Chen, S. Integrated Data and Knowledge Driven Methodology for Human Activity Recognition. *Inf. Sci.* **2020**, *536*, 409–430. [CrossRef]
39. Khaire, P.; Kumar, P.; Imran, J. Combining CNN Streams of RGB-D and Skeletal Data for Human Activity Recognition. *Pattern Recognit. Lett.* **2018**, *115*, 107–116. [CrossRef]
40. Xie, X.; Huang, G.; Zarei, R.; Ji, Z.; Ye, H.; He, J. A Novel Nest-Based Scheduling Method for Mobile Wireless Body Area Networks. *Digit. Commun. Netw.* **2020**. [CrossRef]
41. Zhou, X. Wearable Health Monitoring System based on Human Motion State Recognition. *Comput. Commun.* **2020**, *150*, 62–71. [CrossRef]
42. Li, G.; Li, C. Learning Skeleton Information for Human Action Analysis Using Kinect. *Signal Process. Image Commun.* **2020**, *84*, 115814. [CrossRef]
43. Motion Sensor—Motion, Light and Temperature Sensor. Available online: https://www.fibaro.com/en/products/motion-sensor/ (accessed on 12 August 2020).
44. Technology That's Inventing the Future. Available online: https://www.fitbit.com/us/technology (accessed on 12 August 2020).
45. Mi Band—Understand Your Every Move. Available online: https://www.mi.com/global/miband (accessed on 12 August 2020).
46. MAX25205—Gesture Sensor for Automotive Applications. Available online: https://www.maximintegrated.com/en/products/sensors/MAX25205.html (accessed on 12 August 2020).
47. De, P.; Chatterjee, A.; Rakshit, A. Recognition of Human Behavior for Assisted Living Using Dictionary Learning Approach. *IEEE Sens. J.* **2018**, *16*, 2434–2441. [CrossRef]
48. Lima, Y.; Gardia, A.; Pongsakornsathien, N.; Sabatini, R.; Ezer, N.; Kistan, T. Experimental Characterization of Eye-tracking Sensors for Adaptive Human-Machine Systems. *Measurement* **2019**, *140*, 151–160. [CrossRef]
49. Shu, Y.; Xiong, C.; Fan, S. Interactive Design of Intelligent Machine Vision based on Human–Computer Interaction Mode. *Microprocess. Microsyst.* **2020**, *75*, 103059. [CrossRef]
50. Anitha, G.; Priya, S.B. Posture based Health Monitoring and Unusual Behavior Recognition System for Elderly Using Dynamic Bayesian Network. *Clust. Comput.* **2019**, *22*, 13583–13590. [CrossRef]
51. Wang, Z.; Wang, J.; Zhao, H.; Qiu, S.; Li, J.; Gao, F.; Shi, X. Using Wearable Sensors to Capture Posture of the Human Lumbar Spine in Competitive Swimming. *IEEE Trans. Hum. Mach. Syst.* **2019**, *49*, 194–205. [CrossRef]
52. Concepción, M.A.; Morillo, L.M.S.; García, J.A.A.; González-Abril, L. Mobile Activity Recognition and Fall Detection System for Elderly People Using Ameva Algorithm. *Pervasive Mob. Comput.* **2017**, *34*, 3–13. [CrossRef]
53. Hbali, Y.; Hbali, S.; Ballihi, L.; Sadgal, M. Skeleton-based Human Activity Recognition for Elderly Monitoring Systems. *IET Comput. Vis.* **2018**, *12*, 16–26. [CrossRef]
54. Yu, Z.; Lee, M. Human Motion based Intent Recognition Using a Deep Dynamic Neural Model. *Robot. Auton. Syst.* **2015**, *71*, 134–149. [CrossRef]

55. Bragança, H.; Colonna, J.G.; Lima, W.S.; Souto, E. A Smartphone Lightweight Method for Human Activity Recognition Based on Information Theory. *Sensors* **2018**, *20*, 1856. [CrossRef] [PubMed]
56. Cardenas, E.J.E.; Chavez, G.C. Multimodel Hand Gesture Recognition Combining Temporal and Pose Information based on CNN Descriptors and Histogram of Cumulative Magnitude. *J. Vis. Commun. Image Represent.* **2020**, *71*, 102772. [CrossRef]
57. Lima, W.S.; Souto, E.; El-Khatib, K.; Jalali, R.; Gama, J. Human Activity Recognition Using Inertial Sensors in a Smartphone: An Overview. *Sensors* **2019**, *19*, 3213. [CrossRef] [PubMed]
58. Smedt, Q.D.; Wannous, H.; Vandeborre, J.-P. Heterogeneous Hand Gesture Recognition Using 3D Dynamic Skeletal Data. *Comput. Vis. Image Underst.* **2019**, *181*, 60–72. [CrossRef]
59. Muller, P.; Begin, M.-A.; Schauer, T.; Seel, T. Alignment-Free, Self-Calibrating Elbow Angles Measurement Using Inertial Sensors. *IEEE J. Biomed. Health Inform.* **2017**, *21*, 312–319. [CrossRef]
60. Kianifar, R.; Lee, A.; Raina, S.; Kulic, N. Automated Assessment of Dynamic Knee Valgus and Risk of Knee Injury During the Single Leg Squat. *IEEE J. Transl. Eng. Health Med.* **2017**, *5*, 2100213. [CrossRef]
61. Baghdadi, A.; Cavuoto, L.A.; Crassidis, J.H. Hip and Trunk Kinematics Estimation in Gait Through Kalman Filter Using IMU Data at the Ankle. *IEEE Sens. J.* **2018**, *18*, 4243–4260. [CrossRef]
62. Zhu, C.; Sheng, W. Wearable Sensor-Based Hand Gesture and Daily Activity Recognition for Robot-Assisted Living. *IEEE Trans. Syst. Man Cybern. Part A Syst. Hum.* **2011**, *41*, 569–573. [CrossRef]
63. Guo, Z.; Xiao, F.; Sheng, B.; Fei, H.; Yu, S. WiReader: Adaptive Air Handwriting Recognition Based on Commercial Wi-Fi Signal. *IEEE Internet Things J.* **2020**. [CrossRef]
64. Panwar, M.; Biswas, D.; Bajaj, H.; Jobges, M.; Turk, R.; Maharatna, K.; Acharyya, A. Rehab-Net: Deep Learning Framework for Arm Movement Classification Using Wearable Sensors for Stroke Rehabilitation. *IEEE Trans. Med Eng.* **2019**, *66*, 3026–3037. [CrossRef]
65. Preece, S.J.; Goulermas, J.Y.; Kenney, L.P.J.; Howard, D. A Comparison of Feature Extraction Methods for the Classification of Dynamic Activities from Accelerometer Data. *IEEE Trans. Biomed. Eng.* **2009**, *56*, 871–879. [CrossRef] [PubMed]
66. Zhang, Y.; Chen, K.; Yi, J.; Liu, T.; Pan, Q. Whole-Body Pose Estimation in Human Bicycle Riding Using a Small Set of Wearable Sensors. *IEEE ASME Trans. Mech.* **2016**, *21*, 163–174. [CrossRef]
67. Ross, G.; Dowling, B.; Troje, N.F.; Fischer, S.L.; Graham, R.B. Objectively Differentiating Whole-body Movement Patterns between Elite and Novice Athletes. *Med. Sci. Sports Exerc.* **2018**, *50*, 1457–1464. [CrossRef]
68. Dahmani, D.; Larabi, S. User-Independent System for Sign Language Finger Spelling Recognition. *J. Vis. Commun.* Image Represent. **2014**, *25*, 1240–1250. [CrossRef]
69. Ni, P.; Lv, S.; Zhu, X.; Cao, Q.; Zhang, W. A Light-weight On-line Action Detection with hand Trajectory for Industrial Surveillance. *Digit. Commun. Netw.* **2020**. [CrossRef]
70. Yagi, K.; Sugiura, Y.; Hasegawa, K.; Saito, H. Gait Measurement at Home Using A Single RGB Camera. *Gait Posture* **2020**, *76*, 136–140. [CrossRef] [PubMed]
71. Devanne, M.; Berretti, S.; Pala, P.; Wannous, H.; Daoudi, M.; Bimbo, A.D. Motion Segment Decomposition of RGB-D Sequence of Human Behavior Understanding. *Pattern Recognit.* **2017**, *61*, 222–233. [CrossRef]
72. Xu, W.; Su, P.; Cheung, S.S. Human Body Reshaping and Its Application Using Multiple RGB-D Sensors. *Signal Process. Image Commun.* **2019**, *79*, 71–81. [CrossRef]
73. Wu, H.; Huang, Z.; Hu, B.; Yu, Z.; Li, X.; Gao, M.; Shen, Z. Real-Time Continuous Action Recognition Using Pose Contexts with Depth Sensors. *IEEE Access* **2018**, *6*, 51708. [CrossRef]
74. Ding, C.; Zhang, L.; Gu, C.; Bai, L.; Liao, Z.; Hong, H.; Li, Y.; Zhu, X. Non-Contact Human Motion Recognition based on UWB Radar. *IEEE J. Emerg. Sel. Top. Circuits Syst.* **2018**, *8*, 306–315. [CrossRef]
75. Kim, S.-H.; Geem, Z.W.; Han, G.-T. A Novel Human Respiration Pattern Recognition Using Signals of Ultra-Wideband Radar Sensors. *Sensors* **2019**, *19*, 3340. [CrossRef] [PubMed]
76. Shrestha, A.; Li, H.; Kernec, J.L.; Fioranelli, F. Continuous Human Activity Classification from FMCW Radar with Bi-LSTM Networks. *IEEE Sens. J.* **2020**. [CrossRef]
77. Wang, Y.; Zheng, Y. An FMCW Radar Transceiver Chip for Object Positioning and Human Limb Motion Detection. *IEEE Sens. J.* **2017**, *17*, 236–237. [CrossRef]
78. Han, Z.; Lu, Z.; Wen, X.; Zhao, J.; Guo, L.; Liu, Y. In-Air Handwriting by Passive Gesture Tracking Using Commodity WiFi. *IEEE Commun. Lett.* **2020**. [CrossRef]
79. Li, C.; Liu, M.; Cao, Z. WiHF: Gesture and User Recognition with WiFi. *IEEE Trans. Mob. Comput.* **2020**. [CrossRef]

80. Oguntala, G.A.; Abd-Alhameed, R.A.; Hu, Y.-F.; Noras, J.M.; Eya, N.N.; Elfergani, I.; Rodriguez, J. SmartWall: Novel RFID-Enabled Ambient Human Activity Recognition Using Machine Learning for Unobtrusive Health Monitoring. *IEEE Access* **2019**, *7*, 68022. [CrossRef]
81. Wang, F.; Liu, J.; Gong, W. Multi-Adversarial In-Car Activity Recognition Using RFIDs. *IEEE Trans. Mob. Comput.* **2020**. [CrossRef]
82. Jahanandish, M.H.; Fey, N.P.; Hoyt, K. Lower Limb Motion Estimation Using Ultrasound Imaging: A Framework for Assistive Device Control. *IEEE J. Biomed. Health Inform.* **2019**, *23*, 2505–2514. [CrossRef]
83. Zhou, F.; Li, X.; Wang, Z. Efficient High Cross-User Recognition Rate Ultrasonic Hand Gesture Recognition System. *IEEE Sens. J.* **2020**. [CrossRef]
84. Ling, K.; Dai, H.; Liu, Y.; Liu, A.X. UltraGuest: Fine-Grained gesture sensing and recognition. In Proceedings of the 15th Annual IEEE International Conference on Sensing, Communication, and Networking (SECON), Hong Kong, China, 11–13 June 2018.
85. Vanrell, S.R.; Milone, D.H.; Rufiner, H.L. Assessment of Homomorphic Analysis Human Activity Recognition from Acceleration Signals. *IEEE J. Biomed. Health Inform.* **2018**, *22*, 1001–1010. [CrossRef]
86. Hsu, Y.-L.; Yang, S.-C.; Chang, H.-C.; Lai, H.-C. Human Daily and Sport Activity Recognition Using a Wearable Inertial Sensor Network. *IEEE Access* **2018**, *6*, 31715. [CrossRef]
87. Villeneuve, E.; Harwin, W.; Holderbaum, W.; Janko, B.; Sherratt, R.S. Reconstruction of Angular Kinenatics From Wrist-Worn Inertial Sensor Data for Smart Home Healthcare. *IEEE Access* **2017**, *5*, 2351. [CrossRef]
88. Chelli, A.; Patzold, M. A Machine Learning Approach for Fall Detection and Daily Living Activity Recognition. *IEEE Access* **2019**, *7*, 38670. [CrossRef]
89. Cha, Y.; Kim, H.; Kim, D. Flexible Piezoelectric Sensor-Based Gait Recognition. *Sensors* **2018**, *18*, 468. [CrossRef] [PubMed]
90. Wang, H.; Tong, Y.; Zhao, X.; Tang, Q.; Liu, Y. Flexible, High-Sensitive, and Wearable Strain Sensor based on Organic Crystal for Human Motion Detection. *Org. Electron.* **2018**, *61*, 304–311. [CrossRef]
91. Redd, C.B.; Bamberg, S.J.M. A Wireless Sensory Feedback Device for Real-time Gait Feedback and Training. *IEEE ASME Trans. Mechatron.* **2012**, *17*, 425–433. [CrossRef]
92. Jarrassé, N.; Nicol, C.; Touillet, A.; Richer, F.; Martinet, N.; Paysant, J.; Graaf, J.B. Classification of Phantom Finger, Hand, Wrist, and Elbow Voluntary Gestures in Transhumeral Amputees with sEMG. *IEEE Trans. Neural Syst. Rehabil. Eng.* **2017**, *25*, 68–77. [CrossRef]
93. Li, Z.; Guan, X.; Zou, K.; Xu, C. Estimation of Knee Movement from Surface EMG Using Random Forest with Principal Component Analysis. *Electronics* **2020**, *9*, 43. [CrossRef]
94. Raurale, S.A.; McAllister, J.; Rincon, J.M. Real-Time Embedded EMG Signal Analysis for Wrist-Hand Pose Identification. *IEEE Trans. Signal Process.* **2020**, *68*, 2713–2723. [CrossRef]
95. Di Nardo, F.; Morbidoni, C.; Cucchiarelli, A.; Fioretti, S. Recognition of Gait Phases with a Single Knee Electrogoniometer: A Deep Learning Approach. *Electronics* **2020**, *9*, 355. [CrossRef]
96. Li, H.; Shrestha, A.; Heidari, H.; Kernec, J.L.; Fioranelli, F. Bi-LSTM Network for Multimodel Continuous Human Activity Recognition and Fall Detection. *IEEE Sens. J.* **2020**, *20*, 1191–1201. [CrossRef]
97. Lee, H.-C.; Ke, K.-H. Monitoring of Large-Area IoT Sensors Using a LoRa Wireless Mesh Network System: Design and Evaluation. *IEEE Trans. Instrum. Meas.* **2018**, *67*, 2177–2187. [CrossRef]
98. Kim, J.-Y.; Park, G.; Lee, S.-A.; Nam, Y. Analysis of Machine Learning-based Assessment of Elbow Spasticity Using Inertial Sensors. *Sensors* **2020**, *20*, 1622. [CrossRef] [PubMed]
99. Lee, J.K.; Han, S.J.; Kim, K.; Kim, Y.H.; Lee, S. Wireless Epidermal Six-Axis Inertial Measurement Units for Real-time Joint Angle Estimation. *Appl. Sci.* **2020**, *10*, 2240. [CrossRef]
100. Preece, S.J.; Goulermas, J.Y.; Kenney, L.P.J.; Howard, D.; Meijer, K.; Crompton, R. Activity identification Using Body-mounted Sensors—A Review of Classification Techniques. *Physiol. Meas.* **2009**, *30*, 1–33. [CrossRef]
101. Sun, H.; Lu, Z.; Chen, C.-L.; Cao, J.; Tan, Z. Accurate Human Gesture Sensing with Coarse-grained RF Signatures. *IEEE Access* **2019**, *7*, 81228. [CrossRef]
102. Piris, C.; Gärtner, L.; González, M.A.; Noailly, J.; Stöcker, F.; Schönfelder, M.; Adams, T.; Tassani, S. In-Ear Accelerometer-Based Sensor for Gait Classification. *IEEE Sens. J.* **2020**. [CrossRef]
103. Vu, C.C.; Kim, J. Human Motion Recognition by Textile Sensors based on Machine Learning Algorithms. *Sensors* **2018**, *18*, 3109. [CrossRef]

104. Barut, O.; Zhou, L.; Luo, Y. Multi-task LSTM Model for Human Activity Recognition an Intensity Estimation Using Wearable Sensor Data. *IEEE Internet Things J.* **2020**. [CrossRef]
105. Christ, M.; Braun, N.; Neuffer, J.; Kempa-Liehr, A.W. Time Series FeatuRe Extraction on basis of Scalable Hypothesis Tests (Tsfresh—A Python Package). *Neurocomputing* **2018**, *307*, 72–77. [CrossRef]

Article

People Walking Classification Using Automotive Radar

Linda Senigagliesi *, Gianluca Ciattaglia, Adelmo De Santis and Ennio Gambi

Department of Information Engineering, Università Politecnica delle Marche, via Brecce Bianche 12, 60131 Ancona, Italy; g.ciattaglia@pm.univpm.it (G.C.); adelmo.desantis@univpm.it (A.D.S.); e.gambi@univpm.it (E.G.)

* Correspondence: l.senigagliesi@staff.univpm.it

Received: 24 February 2020; Accepted: 28 March 2020; Published: 30 March 2020

Abstract: Automotive radars are able to guarantee high performances at the expenses of a relatively low cost, and recently their application has been extended to several fields in addition to the original one. In this paper we consider the use of this kind of radars to discriminate different types of people's movements in a real context. To this end, we exploit two different maps obtained from radar, that is, a spectrogram and a range-Doppler map. Through the application of dimensionality reduction methods, such as principal component analysis (PCA) and t-distributed stochastic neighbor embedding (t-SNE) algorithm, and the use of machine learning techniques we prove that is possible to classify with a very good precision people's way of walking even employing commercial devices specifically designed for other purposes.

Keywords: automotive radar; machine learning; walking analysis

1. Introduction

Recognition of a person's type of movement has implications for many aspects of daily life, from security applications to monitoring for assisted living. Discriminating whether a person is running or walking normally in airports or shopping centers, for example, may help video surveillance to detect possible dangerous situations [1–3]. Tools designed for this purpose involve the use of contactless devices, and radar technology is particularly suitable for the mentioned scenario.

Besides its ability to detect the presence of targets of all kinds, sometimes even at considerable distances, radar technology has attracted a large attention thanks to its versatility and usefulness in several fields, from medical applications [4] to traffic surveillance monitoring [5].

In this paper we consider the use of an automotive radar to classify different types of monitored actions. With respect to the work described in Reference [6], the examined activities present less evident differences, since our goal is to distinguish people's way of walking on the basis of their speed. Moreover, the radar here considered works with a higher frequency range and therefore a smaller wavelength, thus allowing a better interaction with objects and improved performance in the micro-Doppler extraction. In addition, the millimeter wave technology exploited allows us to discriminate with a good accuracy also the position of the hands during the walk, whether they are in free movement or hold in pockets. Speed and hands movement classification is performed by using Principal Component Analysis (PCA) and t-distributed Stochastic Neighbor Embedding (t-SNE) methods for feature extraction and supervised learning for the classification task. Different algorithms have been tested, obtaining the best performance in terms of accuracy by using the Nearest Neighbor (NN) and the Support Vector Machine (SVM).

Related Work

Automotive radars with Frequency Modulated Continuous Wave (FMCW) transmission find many applications beyond their original purpose. Recent studies revealed how radars can be exploited to improve road safety and autonomous driving by recognizing the presence of pedestrians starting from micro-Doppler tracks, focusing in particular on the recognition of different parts of the human body [7–9], sometimes applying classification algorithms able to distinguish whether the detected target is a pedestrian or not with a very high accuracy [10]. Through the micro-Doppler components it is also possible to characterize a person's movement or to identify a fall [11]. Moreover, micro-Doppler features have been often exploited in several aspects of human recognition, such as arm motion analysis [12], identification of target human motions [13], or to distinguish people walking in a noisy background [14]. Low power Frequency Modulated Continuous Wave (FMCW) radar and micro-Doppler tracks have been recently used with various scopes, such as discriminating armed from unarmed people [15], identifying people on the basis of their gait characteristics [16–18] or their movements [19], and for gestures recognition [20]. Moreover, radar technology has been successfully applied to the medical field [21,22], for example to remotely monitor the cardiac and respiratory frequency [23]. Principal Component Analysis (PCA) has been often exploited in radar applications, as an instrument to reduce the dimensionality of the available feature space and to automatize the feature extraction and selection procedure [24,25], together with deep learning algorithms for fall detection [26] and human activity recognition [27]. Recent works considered the application of deep learning techniques for gait classification, using smart sensors [28] and radar-based techniques [29] to discriminate aided from unaided motion.

From the year 2022 the 24 GHz bandwidth will no longer be available because of new regulations and ETSI and FCC standards, making it necessary to move towards other frequencies [30]. As regards radar applications, the only available bandwidth will be the ISM bandwidth that have only 250 MHz available, with a performance loss in distance. This explains the presence on the radar market of sensor for industrial and automotive applications working at frequencies over 76 GHz.

As an alternative to radar systems, different contactless technologies have been proposed for gait analysis and walking classification, based on the processing of video (RGB) or video+depth (RGBD) signals. Generally, the main purpose of these research activities is for medical purposes or related to safety issues. In Reference [31], a system able to perform an automatic detection and classification of gait impairments, based on the analysis of a single 2D video, is presented. The main drawback related to the use of video signal is related to privacy issues. To solve this problem, in Reference [32], a representation of gait appearance, with the aim of person identification and classification, is described, based on simple features such as moments extracted from orthogonal view video silhouettes of human walking motion. The availability of a low cost, marker-free, and portable device as Microsoft Kinect Camera allows to develop methods that can respond to the changes in the gait features during the swing stage, tracking the skeletal positional data of body joints to assess and evaluate the gait performance [33]. While being a low cost sensor, Microsoft Kinect is able to track human motion without using wearable sensors and with acceptable reliability. In Reference [34], the standard error of measurement and minimal detectable change sing Kinect is evaluated, confirming the validity of this sensor with standardized clinical tests in individuals with stroke.

The rest of the paper is organized as follows. Section 2 describes the radar used, along with the composition of the transmitted signals and the devices configuration. In Section 3 we outline the signal processing applied to extract spectrograms and maps from the data obtained. Section 4 the dimensionality reduction techniques and classification algorithms applied are introduced. Experimental results are shown in Section 5. Finally Section 6 concludes the paper.

2. Radar System Description

2.1. Used Devices

The radar used in this paper is the Texas Instruments AWR 1642 [35], originally developed for the automotive market [36]. Being designed for industrial applications, it has reduced costs with respect to other types of radar, and its use in the context of interest of this work allows to verify if it is possible to classify people movements with a good accuracy also exploiting commercial devices. It exploits two bandwidths, in the frequency ranges of 76–77 GHz and 77–81 GHz with 1 and 4 GHz bandwidth, respectively. The former is used for long range applications (up to 150 metres) and the latter for short range applications (up to 30 metres). The radar considered in Reference [6] is an Ankortek SDR-Kit 2400AD, a Software Defined Radio (SDR) working in the frequency range 24–26 GHz, with a maximum bandwidth of 2 GHz and a chirp time and maximum ramp slope of 15.625 MHz/μs [37]. With respect to radar used in Reference [6], the higher frequency, the larger bandwidth and the steeper ramp allow to achieve an increase of twenty time in range performance and three times in speed.

An additional characteristic of AWR 1642 is the presence of multiple-input multiple output (MIMO) technology in the sensor [38], which, in case of radar systems, has the goal of improving performance in angular detection.

2.2. Transmitted Signals

Signals are transmitted using Frequency Modulated Continuous Wave (FMCW) modulation, this requires that the transmission frequency varies linearly from a minimum value to a maximum value in a time interval, called *chirp* time. The transmitted chirps are grouped into frames. Inside each frame, which has a time duration called *periodicity*, the radar transmits a certain number of chirps, as schematized in Figure 1.

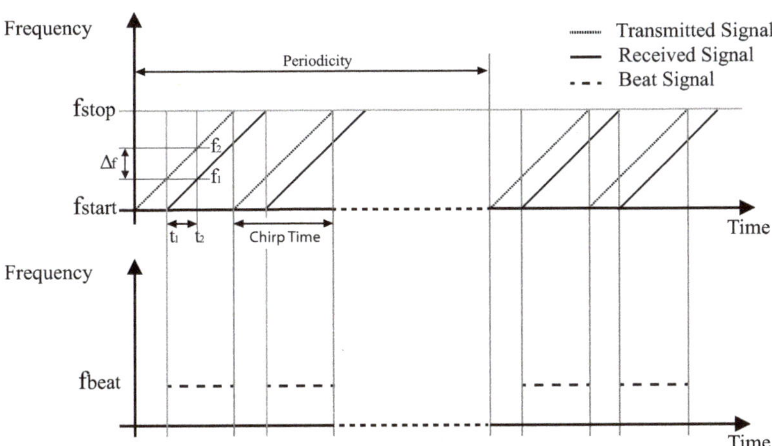

Figure 1. RADAR frame periodicity.

Each chirp is built as shown in Figure 2. We have an *Idle Time*, needed because the ramp generator requires some time to restart the ramp and generate a new chirp. Then a guard time, or analog-to-digital converter (ADC) Valid Start Time, is considered in the first part of the ramp, which is not linear and may lead to a performance reduction, as described in Reference [39].

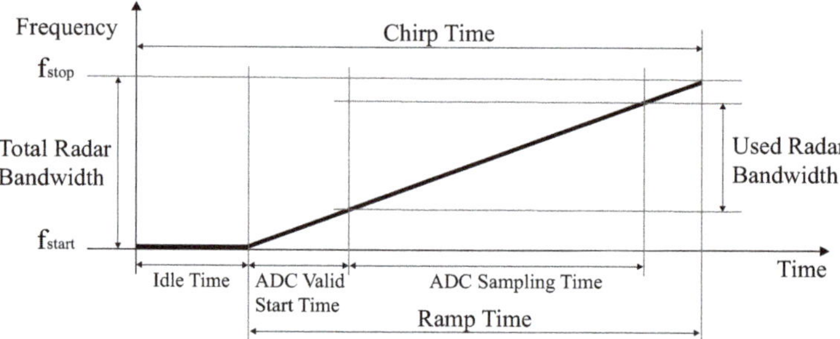

Figure 2. Chirps timing.

Then we have the effective ADC Sampling Time, which represents the time duration of the ramp acquired by the radar. Within this interval analog-to-digital converters (ADCs) samples of the IF signals are collected. As easily observable in Figure 2, a time shorter than the total ramp time t_{ramp} is used, so the used radar bandwidth B is smaller than the maximum possible TB, and is calculated as

$$B = ADC_{SamplingTime} \cdot S \leq TB = t_{ramp} \cdot S, \quad (1)$$

where S represents the slope of the ramp and $ADC_{SamplingTime}$ is given by the product between the the number of samples $n_{samples}$ acquired for each chirp and the sampling period $t_{sampling}$. The devices configuration must take in account these parameters in order to avoid as much as possible the non linear effects of the sensor.

The importance of avoiding the first part of the ramp is evident from the analysis of the intermediate frequency (IF) signal during time and on the complex plane. As briefly described in Reference [40], it is possible to see this effect on the IQ plot. Using different calibrations, the first with analog-to-digital converter (ADC) Valid Start Time equal to zero, and the second with time equal to 6 µs it is possible to observe how this imperfection can be avoided without the need of using an algorithm. Figure 3a shows 500 samples of IF signal across two different chirps. If the guard time is not used, there is a spike at the beginning of the first chirp. In Figure 3b the case of the same segment of IF signal, with an analog-to-digital converter (ADC) Valid Start Time of 6 µs is depicted, not showing the same effect. The spike disappears with a minimum value of 3 µs.

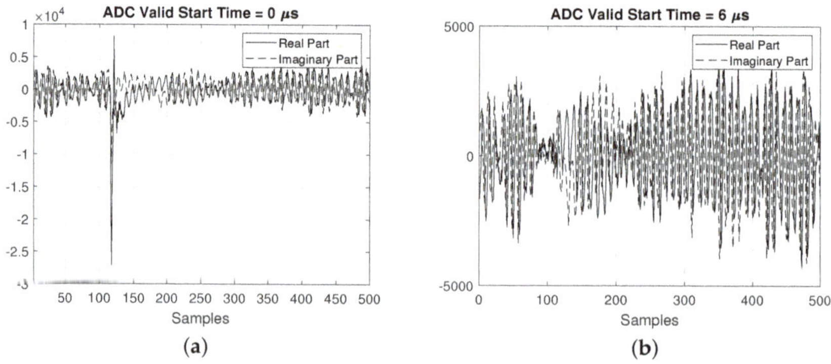

Figure 3. IF signal (**a**) without guard interval, highlighting the presence of a spike, and (**b**) with guard interval.

3. Radar Signal Processing

On the basis of the used configuration we have four available receiving lines. To perform our analysis we need just one of them, so we can sum the complex samples coming from the analog-to-digital converters in order to improve the signal-to-noise ratio. We thus obtain a vector of complex samples which can be reorganized in the form of matrix, as shown in Figure 4. Along the rows of the matrix, also called *fast time*, we have samples from single chirps, while on the columns, or *slow time*, we samples from different chirps.

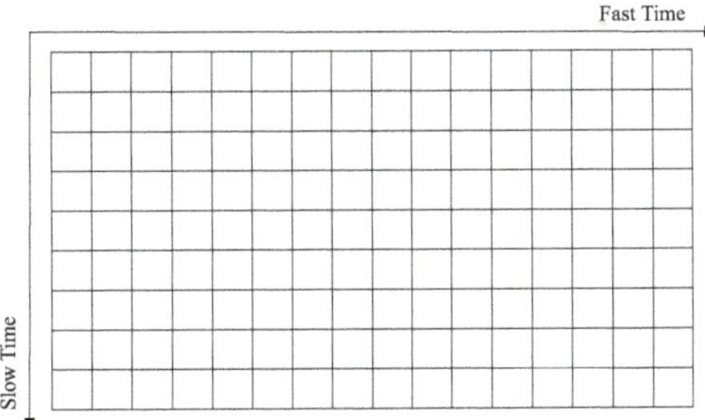

Figure 4. Slow Time and Fast Time Matrix.

From this raw matrix we can extract two types of map to classify different types of movement. The first one contains information about distance and speed of the subject and it is obtained by applying a Fourier transform to columns and then to rows; this map is defined as *Range-Doppler Map*. Each element of the map is a complex number and it is built considering the total acquisition, as shown in Figure 5. Besides distance and speed, this allows to extract also the micro-Doppler components characterizing the movement.

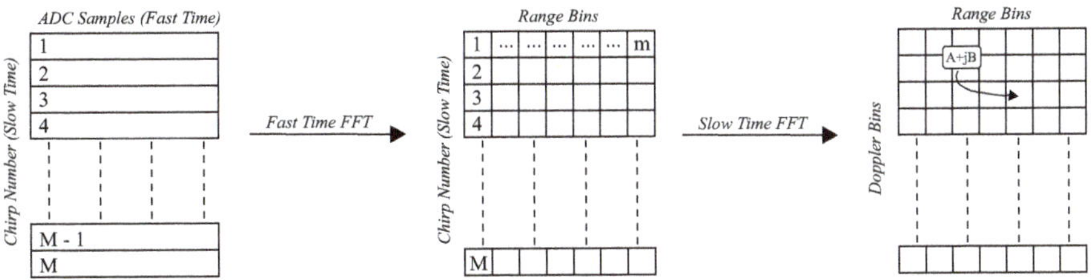

Figure 5. Range-Doppler Map data processing.

Since our subjects are moving during the acquisitions, we can extract the spectrogram from each Range bin and hence characterize their micro-Dopplers along the entire activity.

During each acquisition all the objects are stationary with the exception of the subject under exam, therefore the only significant micro-Doppler components are related to her/him. The presence of stationary objects does not influence either the Range-Doppler map, but only the zero Doppler. As described in Reference [19], from this map is possible to identify the kind of movement carried out by the subject.

The second type of map can be extracted through spectrograms and is denoted as *Doppler-Time Map*. A spectrogram is the most common time-frequency representation [41], and it is derived from the Short Time Fourier Transform (STFT) according to the following equation

$$STFT_x[k,n] = \sum_{r=-\infty}^{+\infty} x[r]w[n-r]e^{-j2\pi rk/N},\ k = 0,1,\cdots,N-1, \tag{2}$$

where n represents a discrete index of time, k is a discrete index of frequency and $w[\cdot]$ is a window function. The Short Time Fourier Transform (STFT) can in fact be considered as the Fourier transform of a signal multiplied by a window sliding over time. A trade-off between resolution in time and in frequency must be found, and overlapping frequencies can help in this sense [42].

Starting from the range matrix, the second matrix in Figure 5, and applying the Short Time Fourier Transform (STFT) along the rows, we obtain the Doppler-Time map. This function uses windows of 512 samples, with an overlap of 98% and an Hann window is applied. Figure 6 depicts this process.

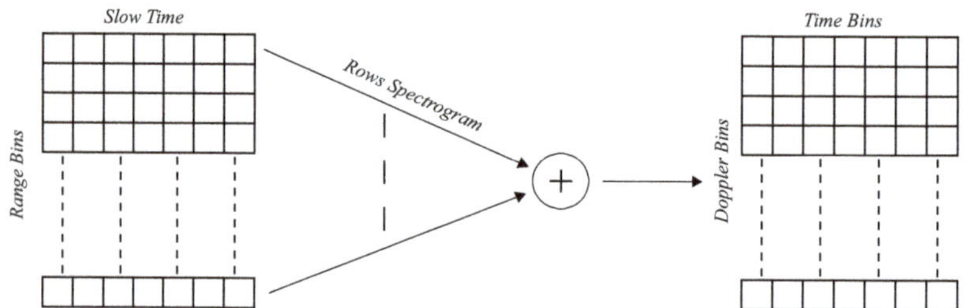

Figure 6. Doppler-Time Spectrograms.

By using both the mentioned maps it is possible to classify different kinds of movements, as will be explained in the next section.

4. Movements Classification

In this section we briefly describe the dimensionality reduction techniques and the classification algorithms used in the following section to discriminate the kinds of activities under consideration. As regards features extraction, we resort to two different methods to reduce data dimensionality, the Principal Component Analysis (PCA) and the t-distributed Stochastic Neighbor Embedding (t-SNE) .

Both the maps obtained through the radar signal processing, that is, the Range-Doppler map and Time-Doppler map, are considered as images. Vectors resulting from the application of dimensionality reduction techniques to these images, that is, the principal components extracted from Principal Component Analysis (PCA) and the main dimensions given by the t-distributed Stochastic Neighbor Embedding (t-SNE) , will serve as features vectors. We have a set of N images I_n of dimension $[l \times m]$, with $n = 1, \cdots, N$. Images are initially vectorized row-wise and grouped in order to form a training set $\mathbf{X} = [x^{(1)}, \cdots, x^{(N)}]^T$, where T denotes the transpose operator; rows of $\mathbf{X}$ correspond to observations and columns correspond to variables.

4.1. *Principal Component Analysis*

Principal Component Analysis (PCA) [43] is a non supervised transform also known as Karhunen-Loeve transform (KLT). It aims at finding suitable linear transformations y of the observed variables that are easily interpreted and capable of highlighting and summarizing the information inherent in the initial matrix I. This tool is especially useful when dealing with a considerable number of variables from which you want to extract the greatest possible information while working with a smaller set of variables.

Principal Component Analysis (PCA) can hence be described as a transformation of a given set of N vectors into inputs (variables) of the same length K placed in a vector N-dimensional $\mathbf{X}$, which allows to transform this vector into a second vector y, built in such a way that the first element of y includes the greatest possible variability (and therefore more information) of the original variables, that the second represents the greater variability of the x_i after the first component, and so up to $y^{(N)}$ which takes into account the smallest fraction of the original variance. Therefore the main components are those linear combinations of the random variables $x^{(N)}$ according to the unit norm which make the variance maximum and which are uncorrelated.

The resulting vector y form the feature vector which can be used for classification. Moreover, Principal Component Analysis (PCA) algorithm is invertible, so original data can be easily recovered.

4.2. t-distributed Stochastic Neighbor Embedding

t-distributed Stochastic Neighbor Embedding (t-SNE) [44] is a non linear and non supervised transform, specifically designed to reduce dimensionality to 2 or 3 dimensions in order to display multidimensional data.

The t-distributed Stochastic Neighbor Embedding (t-SNE) algorithm consists of two main steps. Given our set of N vectorized images $x^{(1)}, \cdots, x^{(N)}$ with length $l \cdot m$, t-SNE first computes the conditional probability $p_{j|i}$, which represents the similarity of datapoint x_j to datapoint x_i. In other words, it evaluates the probability that x_i would pick x_j as its neighbor if neighbors were picked in proportion to their probability density under a Gaussian centered at x_i. In formulas,

$$p_{j|i} = \frac{\exp\left(-\|x_i - x_j\|^2 / 2\sigma_i^2\right)}{\sum_{k \neq i} \exp\left(-\|x_i - x_j\|^2 / 2\sigma_i^2\right)}. \tag{3}$$

t-distributed Stochastic Neighbor Embedding (t-SNE) then defines a similar probability distribution over the points in the low-dimensional map, and it minimizes the Kullback-Leibler (KL) divergence between the two distributions with respect to the locations of the points in the map.

4.3. Classification Algorithms

As regards the classification task, we consider the use of k-Nearest Neighbor (NN) and Support Vector Machines (SVMs). They are both supervised and non parametric algorithms.

k-Nearest Neighbor (NN) is an instance-based algorithm, meaning that it does not explicitly learn a model. Instead, it chooses to memorize the training instances which are subsequently used as "knowledge" for the forecasting phase. In concrete terms, this means that only when a query is made in the database (i.e., when asked to provide a label with an input), the algorithm will use the training instances to send a response. As a drawback, this algorithm presents both a storage cost during the training phase, since it is necessary to store a potentially huge dataset, and a computational cost during the prediction phase since the classification of a given observation requires the vision and/or analysis of the entire dataset. In the context of classification, the k-Nearest Neighbor (NN) algorithm essentially boils down to determine a majority vote among the k closest neighbors to a given unknown instance. The proximity is defined by a distance metric, usually the Euclidean distance, between two data points.

Support Vector Machine (SVM) algorithm classifies data by creating a linear of non-linear decision boundary to separate different classes. It projects the data through a non-linear function to a space with a higher dimension, lifting them from their original space to a feature space, which can be of unlimited dimension. To perform this operation, Support Vector Machine (SVM) makes use of kernels, among which one of the most used is the Gaussian kernel.

5. Experimental Results

5.1. Experimental Setup

In our experiments we consider a data set composed by nineteen subjects who repeated each activity for three times, for a total of 168 different acquisitions. Three different activities were examined:

- Slow walk;
- Fast walk;
- Slow walk with hands in pockets.

Note that we do not consider a data set built ad hoc: each subject was simply asked to walk in a "slow" or in a "fast" way, without specifying the number of steps or the time required to complete the activity, in order to generate data as realistic as possible. In addition acquisitions belonging to subjects of different height and weight were collected to provide a set comprehensive of a large variety of characteristics. Walking speed difference is subtle and depends on the person examined, who interpreted it subjectively. In general, the average speed measured for the fast walk is around 2 m/s, while for the slow walk, with both free hands or hands in pockets is about 1.2 m/s. Differences in subjects' speed, including the "holding the arm" case (which is similar to our "hands in pockets"), have been considered in References [45,46], although their datasets were composed by 8 and 3 subjects, respectively.

The radar configuration parameters are chosen according to the measurement area selected and to the kind of activity required to the subjects. Some parameters are chosen according to the following range R equation

$$R = \frac{f_{beat} \cdot c}{2 \cdot S}, \tag{4}$$

where f_{beat} is the beat frequency and c is the speed of light. We can evaluate the maximum speed of the target as

$$v_{target} = \frac{\lambda}{4 \cdot t_{chirp}}, \tag{5}$$

where λ represents the wavelength of the transmitted signal and t_{chirp} is the time duration of the chirp. The measurement area is an hallway, about 12 meters long, and is free of furniture. During each activity the subject goes from the starting point in front of the RADAR to a distance of about 9 meters, and then comes back. Due the fact that the measurement time is of 16 seconds, it is possible that the acquisition ends before the subject returns to the initial position. The parameters used for the measurements are reported in Table 1.

A first analysis has been made on the background without any subject, which is depicted in Figure 7. Only one measurement has been performed, since the test area is the same for all the subjects. From this analysis is possible to see that the background does not affect the measurements, thus we can neglect its effect in the movements classification.

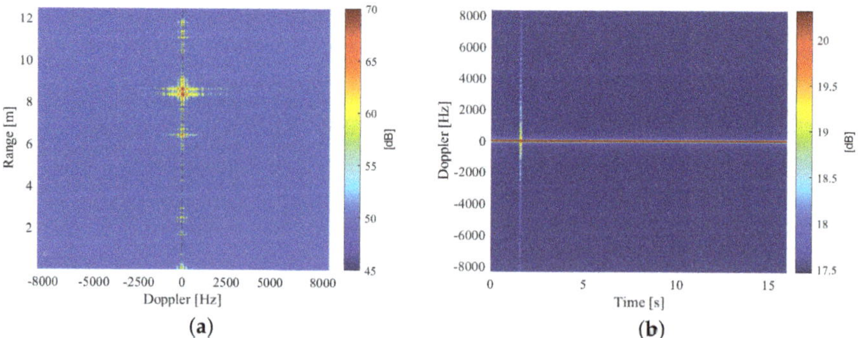

Figure 7. Analysis on the background in absence of subjects using (**a**) Range-Doppler map and (**b**) Doppler-Time map.

In Figure 8 we show an example a subject walking in different ways, displaying both Range-Doppler maps (on the left) and Doppler-time maps (on the right). It is possible to observe that slow and fast walk are easily recognizable in the maps. As expected, maps related to slow walk with hands in pockets present a slightly less evident Doppler with respect to free hands, but this effect is scarcely noticeable.

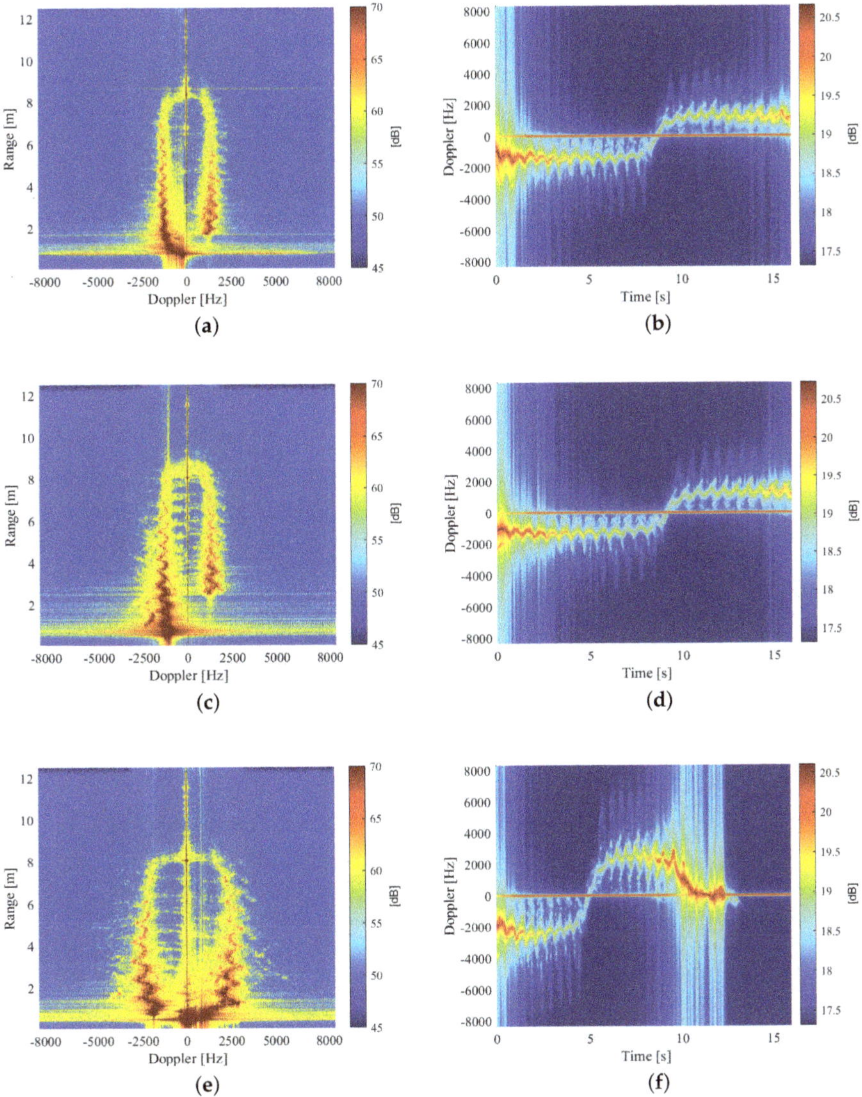

Figure 8. Example of a person walking slowly (**a**,**b**), slowly with hands in pockets (**c**,**d**) and fast (**e**,**f**).

Table 1. RADAR parameters.

Parameter	Value
f_{start}	77 GHz
S	60.012 MHz/µs
t_{idle}	100 µs
ADC Valid Start Time	6 µs
f_s	10 Msps
t_{ramp}	60 µs
$n_{samples}$	512
n_{frame}	400
no. of chirps per frame	128
Periodicity	40 ms
Used Radar Bandwidth	3.6 GHz

5.2. Classification

Data obtained after the processing of the radar signal are treated as images. Since their original size cannot be easily handled, all matrices have been reshaped to the same dimension $[195 \times 119]$. In order to further reduce dimensionality and to extract features from images, Principal Component Analysis (PCA) and t-distributed Stochastic Neighbor Embedding (t-SNE) algorithm have been then applied separately to data.

In Figure 9a,b we show the classification accuracy resulting from exploiting a different number of principal components, by using a Nearest Neighbor (NN) classifier and a Support Vector Machine (SVM) algorithm. We choose to use a Gaussian kernel for the Support Vector Machine (SVM).The value of k for the k-Nearest Neighbor (NN) and the kernel used for Support Vector Machine (SVM) have been chosen by using a leave-one-out cross-validation algorithm, which aims at minimizing the validation error. Each sample of the dataset is alternatively selected as a validation set, whilst the remaining part represents the training set. In this way all samples are used only one time both for training, both for validation. Results obtained by the algorithm for odd values of k between 1 and 49 are shown in Figure 10, where k equal to 1 leads to an error of about 2.4%. The validation error obtained by different kernels in percentage is reported in Table 2, thus directing the choice to the use of linear kernel in our scenario.

Table 2. Results of the leave-one-out cross validation for support vector machine (SVM) with different kernels.

Kernel	Linear	Gaussian	Polynomial
Error validation (%)	4.46	17.26	33.33

Sixty percent of the acquisitions are used for training, while the remainder is used for testing. Results have been averaged over 100 classification results obtained choosing training and test sets at random. We consider here only two classes, corresponding to the slow and fast walk. Interestingly, it is possible to observe that the number of principal components (or number of dimension in case of t-distributed Stochastic Neighbor Embedding (t-SNE)) that here corresponds to the number of features, has a small impact on the classification performance. The application of Principal Component Analysis (PCA) or t-distributed Stochastic Neighbor Embedding (t-SNE) algorithm to extract features from images leads to very similar results, although t-distributed Stochastic Neighbor Embedding (t-SNE) was originally designed to reduce data to two or three dimensions, and becomes very slow for higher values. In addition, we obtain the same results using both Range-Doppler and Doppler-Time maps.

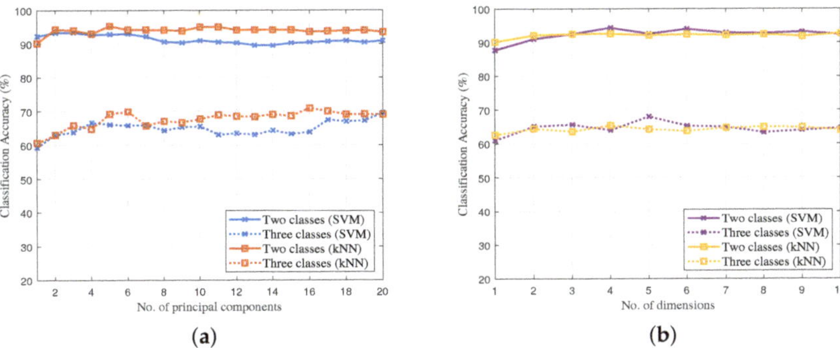

Figure 9. Comparison of classification accuracy achieved by SVM and kNN considering 2 and 3 classes, applying (**a**) Principal Component Analysis (PCA) and (**b**) t-distributed stochastic neighbor embedding (t-SNE).

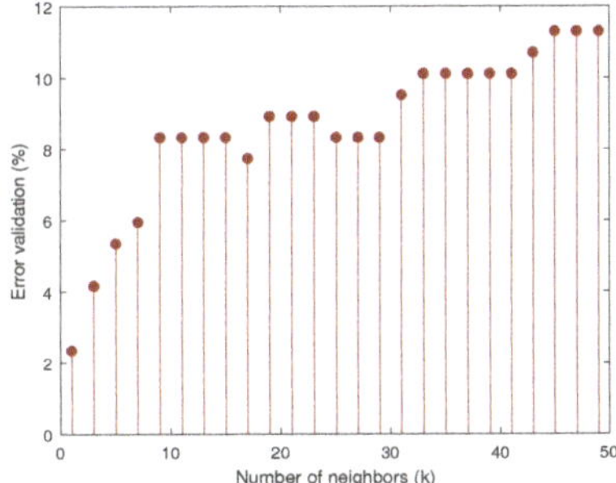

Figure 10. Results of the leave-one-out cross-validation for the k-Nearest Neighbor (kNN).

In Tables 3 and 4 we show the confusion matrices obtained by applying classification on two and three different classes. In the first table, measurements of slow walk and slow walk with hands in pockets have been incorporated into a single class, while in the second table they have been split into two separate classes. As predictable, distinguishing free hands from hands in pockets is a much more complicated task than identify different ways of walking. In the first case in fact the best accuracy obtained is about 72% and red boxes highlight the presence of a number of misclassified examples, although the fast walk is recognized from the other activities with a high precision (87.5%); SVM methods seem to achieve better performance than KNN algorithms. In the latter case we have instead an excellent accuracy of more than 93%. In both Tables 3 and 4 we highlighted a high presence of correct detections in green, while a high number of misclassified samples is marked in red.

Table 3. Confusion matrix obtained applying SVM and kNN (into parentheses) on two classes, considering 5 principal components, $acc = 93.5\%$.

True/Predicted	S	F
Slow Walk (S)	110 (109)	2 (3)
Fast Walk (F)	9 (8)	47 (48)

Table 4. Confusion matrix obtained applying SVM and kNN (into parentheses) on three different classes, considering 9 principal components, $acc_{SVM} = 72\%$, $acc_{KNN} = 66.7\%$.

True/Predicted	S	F	SH
Slow Walk (S)	33 (32)	2 (1)	21 (23)
Fast Walk (F)	4 (5)	49 (48)	3 (3)
Slow Walk with Hands in Pockets (SH)	16 (22)	1 (2)	39 (32)

In Table 5, we give an overview of the results obtained by other works focused on the classification of walking activities through radar measurements, showing the best accuracies achieved. [*] denotes the present work. In Reference [45] 7 types of activities are considered, that is, walking backwards, limping, depressed, elderly, excited, holding the arm and walking in a zigzag, and the radar used is an Ultra-Wide Band; Reference [46] considers a Frequency Modulated Continuous Wave (FMCW) radar, and the examined activities are crawl, creep on hands and knees, walk, jog and run. Although the difference between walking slowly or quickly is less evident than the other activities, we prove that our system is able to achieve a better accuracy. Moreover, we consider a larger number of subjects that move differently from each other, thus confirming the validity of our method in a realistic context. The activity of holding the arm while walking [45], which is in some way comparable to our case of walking slowly with hands in pockets, could not be differentiated from the others at all, with a specific accuracy of 42.42% (see Reference [45], Table 2).

Table 5. Comparison of different radar based methods for human walking classification.

	Radar Type	No. of Activities	Dataset Dimension	Algorithm	Best Accuracy
[*]	FMCW mmWave	2	19 subjects, 168 acquisitions	PCA/t-SNE + k-NN/SVM	93.5%
[*]	FMCW mmWave	3	19 subjects, 168 acquisitions	PCA/t-SNE + k-NN/SVM	72%
[45]	Ultra Wide Band	7	8 subjects, 280 acquisitions	PCA + SVM	89.1%
[46]	FMCW mmWave	5	3 subjects, 95 acquisitions	CV/TV + SVM	91%

The subjectivity and the personal speed interpretation of the conducted tests represents the major error source for our classification model. A standardized time or number of steps during the experiment should probably improve the system performance, but this would not represent a realistic scenario and it is out of the scope of this work.

6. Conclusions and Future Works

We have assessed the performance of an automotive radar to classify different types of movements, focusing our attention to the distinction of people's way of walking. The dataset was not built ad hoc, but we have collected acquisitions of subjects with different characteristics free to walk in a given indoor environment. We have considered the use of PCA and t-SNE techniques to reduce data dimensionality and to extract features, and then we have applied different classification algorithms. From the obtained results it is possible to state that movement classification of human targets is a much more complex task with respect to the discrimination of people from other objects. However, we have shown that, by exploiting the micro-Doppler components of the radar signal, we are able to identify with a high accuracy slow and fast walking. We have also characterized the presence or absence of movement of the arms with more than 72% of precision, which represents a good starting point for a future work. A possible future direction may also include the investigation of deep learning methods in our scenario in order to better distinguish small movements.

Author Contributions: G.C. designed the system, G.C. and L.S. performed the experimental tests and data processing, writing also the main part of the paper. A.D.S. and E.G. participated in data collection and processing. E.G. coordinated the project, the discussion of result, and the manuscript writing. All authors have read and agreed to the published version of the manuscript.

Funding: This research received no external funding.

Conflicts of Interest: The authors declare no conflict of interest.

Abbreviations

The following abbreviations are used in this manuscript:

ADC	analog-to-digital converter
FMCW	Frequency Modulated Continuous Wave
MIMO	multiple-input multiple output
NN	Nearest Neighbor
PCA	Principal Component Analysis
SDR	Software Defined Radio
STFT	Short Time Fourier Transform
SVM	Support Vector Machine
t-SNE	t-distributed Stochastic Neighbor Embedding

References

1. Frazier, L.M. MDR for law enforcement [motion detector radar]. *IEEE Potentials* **1998**, *16*, 23–26. [CrossRef]
2. Mazel, D.S.; Barry, A. Mobile Ravin: Intrusion Detection and Tracking with Organic Airport Radar and Video Systems. In Proceedings of the 40th Annual 2006 International Carnahan Conference on Security Technology, Lexington, KY, USA, 16–19 October 2006; pp. 30–33. [CrossRef]
3. Zyczkowski, M.; Palka, N.; Trzcinski, T.; Dulski, R.; Kastek, M.; Trzaskawka, P. Integrated radar-camera security system: Experimental results. In Proceedings of the Radar Sensor Technology XV, Orlando, FL, USA, 25–29 April 2011; Volume 8021, p. 80211U.
4. Pisa, S.; Pittella, E.; Piuzzi, E. A survey of radar systems for medical applications. *IEEE Aerosp. Electron. Syst. Mag.* **2016**, *31*, 64–81. [CrossRef]
5. Roy, A.; Gale, N.; Hong, L. Automated traffic surveillance using fusion of Doppler radar and video information. *Math. Comput. Model.* **2011**, *54*, 531–543. [CrossRef]
6. Seifert, A.K.; Schäfer, L.; Amin, M.G.; Zoubir, A.M. Subspace Classification of Human Gait Using Radar Micro-Doppler Signatures. In Proceedings of the 2018 26th European Signal Processing Conference (EUSIPCO), Rome, Italy, 3–7 September 2018; pp. 311–315. [CrossRef]
7. Chen, V.C. Doppler signatures of radar backscattering from objects with micro-motions. *IET Signal Process.* **2008**, *2*, 291–300. [CrossRef]
8. Tahmoush, D. Review of micro-Doppler signatures. *IEE Proc.-Radar Sonar Navig.* **2015**, *9*, 1140–1146. [CrossRef]
9. Held, P.; Steinhauser, D.; Kamann, A.; Holdgrün, T.; Doric, I.; Koch, A.; Brandmeier, T. Radar-Based Analysis of Pedestrian Micro-Doppler Signatures Using Motion Capture Sensors. In Proceedings of the 2018 IEEE Intelligent Vehicles Symposium (IV), Changshu, China, 26–30 June 2018; pp. 787–793. [CrossRef]
10. Prophet, R.; Hoffmann, M.; Vossiek, M.; Sturm, C.; Ossowska, A.; Malik, W.; Lübbert, U. Pedestrian Classification with a 79 GHz Automotive Radar Sensor. In Proceedings of the 2018 19th International Radar Symposium (IRS), Bonn, Germany, 20–22 June 2018; pp. 1–6. [CrossRef]
11. Cippitelli, E.; Fioranelli, F.; Gambi, E.; Spinsante, S. Radar and RGB-Depth Sensors for Fall Detection: A Review. *IEEE Sens. J.* **2017**, *17*, 3585–3604. [CrossRef]
12. Zeng, Z.; Amin, M.G.; Shan, T. Arm Motion Classification Using Time-Series Analysis of the Spectrogram Frequency Envelopes. *Remote Sens.* **2020**, *12*, 454. [CrossRef]

13. Ma, X.; Zhao, R.; Liu, X.; Kuang, H.; Al-qaness, M.A. Classification of Human Motions Using Micro-Doppler Radar in the Environments with Micro-Motion Interference. *Sensors* **2019**, *19*, 2598. [CrossRef]
14. Kwon, J.; Kwak, N. Radar Application: Stacking Multiple Classifiers for Human Walking Detection Using Micro-Doppler Signals. *Appl. Sci.* **2019**, *9*, 3534. [CrossRef]
15. Fioranelli, F.; Ritchie, M.; Griffiths, H. Analysis of polarimetric multistatic human micro-Doppler classification of armed/unarmed personnel. In Proceedings of the 2015 IEEE Radar Conference (RadarCon), Arlington, VA, USA, 10–15 May 2015; pp. 0432–0437. [CrossRef]
16. Ricci, R.; Balleri, A. Recognition of humans based on radar micro-Doppler shape spectrum features. *IEE Proc.-Radar Sonar Navig.* **2015**, *9*, 1216–1223. [CrossRef]
17. Fioranelli, F.; Ritchie, M.; Griffiths, H. Performance Analysis of Centroid and SVD Features for Personnel Recognition Using Multistatic Micro-Doppler. *IEEE Geosci. Remote Sens. Lett.* **2016**, *13*, 725–729. [CrossRef]
18. Vandersmissen, B.; Knudde, N.; Jalalvand, A.; Couckuyt, I.; Bourdoux, A.; De Neve, W.; Dhaene, T. Indoor Person Identification Using a Low-Power FMCW Radar. *IEEE Geosci. Remote Sens. Lett.* **2018**, *56*, 3941–3952. [CrossRef]
19. Gambi, E.; Ciattaglia, G.; De Santis, A. People Movement Analysis with Automotive Radar. In Proceedings of the 2019 IEEE 23rd International Symposium on Consumer Technologies (ISCT), Ancona, Italy, 19–21 June 2019; pp. 233–237. [CrossRef]
20. Dekker, B.; Jacobs, S.; Kossen, A.S.; Kruithof, M.C.; Huizing, A.G.; Geurts, M. Gesture recognition with a low power FMCW radar and a deep convolutional neural network. In Proceedings of the 2017 European Radar Conference (EURAD), Nuremberg, Germany, 11–13 October 2017; pp. 163–166. [CrossRef]
21. Anishchenko, L.; Alekhin, M.; Tataraidze, A.; Ivashov, S.; Bugaev, A.S.; Soldovieri, F. Application of step-frequency radars in medicine. In Proceedings of the Radar Sensor Technology XVIII, Baltimore, MD, USA, 5–9 May 2014; Volume 9077, pp. 524–530. [CrossRef]
22. Rissacher, D.; Galy, D. Cardiac radar for biometric identification using nearest neighbour of continuous wavelet transform peaks. In Proceedings of the IEEE International Conference on Identity, Security and Behavior Analysis (ISBA 2015), Hong Kong, China, 23–25 March 2015; pp. 1–6. [CrossRef]
23. Ciattaglia, G.; Senigagliesi, L.; De Santis, A.; Ricciuti, M. Contactless measurement of physiological parameters. In Proceedings of the 2019 IEEE 9th International Conference on Consumer Electronics (ICCE-Berlin), Berlin, Germany, 8–11 September 2019; pp. 22–26. [CrossRef]
24. Zabalza, J.; Clemente, C.; Di Caterina, G.; Ren, J.; Soraghan, J.J.; Marshall, S. Robust PCA micro-doppler classification using SVM on embedded systems. *IEEE Trans. Aerosp. Electron. Syst.* **2014**, *50*, 2304–2310. [CrossRef]
25. Jokanović, B.; Amin, M.; Ahmad, F.; Boashash, B. Radar fall detection using principal component analysis. In Proceedings of the Radar Sensor Technology XX, Baltimore, MD, USA, 17–21 April 2016; Volume 9829, p. 982916.
26. Jokanovic, B.; Amin, M.; Ahmad, F. Radar fall motion detection using deep learning. In Proceedings of the 2016 IEEE Radar Conference (RadarConf), Philadelphia, PA, USA, 2–6 May 2016; pp. 1–6. [CrossRef]
27. Li, X.; He, Y.; Jing, X. A Survey of Deep Learning-Based Human Activity Recognition in Radar. *Remote Sens.* **2019**, *11*, 1068. [CrossRef]
28. Lee, S.S.; Choi, S.T.; Choi, S.I. Classification of gait type based on deep learning using various sensors with smart insole. *Sensors* **2019**, *19*, 1757. [CrossRef]
29. Seyfioğlu, M.S.; Gürbüz, S.Z.; Özbayoğlu, A.M.; Yüksel, M. Deep learning of micro-Doppler features for aided and unaided gait recognition. In Proceedings of the 2017 IEEE Radar Conference (RadarConf), Seattle, WA, USA, 8–12 May 2017; pp. 1125–1130. [CrossRef]
30. Ramasubramanian, K.; Ramaiah, K.; Aginskiy, A. Moving from Legacy 24 GHz to State-of-the-Art 77 GHz Radar. Available online: http://www.ti.com/lit/wp/spry312/spry312.pdf (accessed on 30 March 2020).
31. Verlekar, T.T.; Soares, L.D.; Correia, P.L. Automatic Classification of Gait Impairments Using a Markerless 2D Video-Based System. *Sensors* **2018**, *18*. [CrossRef]
32. Lee, L.; Grimson, W.E.L. Gait analysis for recognition and classification. In Proceedings of the Fifth IEEE International Conference on Automatic Face Gesture Recognition, Washington, DC, USA, 21–21 May 2002; pp. 155–162. [CrossRef]
33. Elkurdi, A.; Soufian, M.; Nefti-Meziani, S. Gait speeds classifications by supervised modulation-based machine-learning using kinect camera. *J. Med. Res. Innov.* **2018**, *2*, 1–6. [CrossRef]

34. Latorre, J.; Colomer, C.; Alcañiz, M.; Llorens, R. Gait analysis with the Kinect v2: Normative study with healthy individuals and comprehensive study of its sensitivity, validity, and reliability in individuals with stroke. *J. Neuroeng. Rehabil.* **2019**, *16*, 97. [CrossRef]
35. Instruments, T. AWR1642 Single-Chip 77- and 79-GHz FMCW Radar Sensor. Available online: http://www.ti.com/lit/ds/symlink/awr1642.pdf (accessed on 30 March 2020).
36. Patole, S.M.; Torlak, M.; Wang, D.; Ali, M. Automotive radars: A review of signal processing techniques. *IEEE Signal Process. Mag.* **2017**, *34*, 22–35. [CrossRef]
37. Ancortek. SDR 2400AD. Available online: http://ancortek.com/wp-content/uploads/2019/04/SDR-2400AD-Datasheet.pdf (accessed on 30 March 2020).
38. Li, J.; Stoica, P. MIMO Radar with Colocated Antennas. *IEEE Signal Process. Mag.* **2007**, *24*, 106–114. [CrossRef]
39. Brennan, P.; Huang, Y.; Ash, M.; Chetty, K. Determination of sweep linearity requirements in FMCW radar systems based on simple voltage-controlled oscillator sources. *IEEE Trans. Aerosp. Electron. Syst.* **2011**, *47*, 1594–1604. [CrossRef]
40. Alizadeh, M.; Shaker, G.; De Almeida, J.C.M.; Morita, P.P.; Safavi-Naeini, S. Remote monitoring of human vital signs using mm-wave FMCW radar. *IEEE Access* **2019**, *7*, 54958–54968. [CrossRef]
41. Mitra, S.K.; Kuo, Y. *Digital Signal Processing: A Computer-Based Approach*; McGraw-Hill: New York, NY, USA, 2006; Volume 2.
42. Cohen, L. *Time-Frequency Analysis*; Prentice Hall: Upper Saddle River, NJ, USA, 1995; Volume 778.
43. Wold, S.; Esbensen, K.; Geladi, P. Principal component analysis. *Chemom. Intell. Lab.* **1987**, 2, 37–52. [CrossRef]
44. Maaten, L.V.D.; Hinton, G. Visualizing data using t-SNE. *J. Mach. Learn. Res.* **2008**, *9*, 2579–2605.
45. Bryan, J.D.; Kwon, J.; Lee, N.; Kim, Y. Application of ultra-wide band radar for classification of human activities. *IEE Proc.-Radar Sonar Navig.* **2012**, *6*, 172–179. [CrossRef]
46. Björklund, S.; Petersson, H.; Hendeby, G. Features for micro-Doppler based activity classification. *IEE Proc.-Radar Sonar Navig.* **2015**, *9*, 1181–1187. [CrossRef]

Article

Closing the Wearable Gap—Part III: Use of Stretch Sensors in Detecting Ankle Joint Kinematics During Unexpected and Expected Slip and Trip Perturbations

Harish Chander [1,*], Ethan Stewart [1], David Saucier [2], Phuoc Nguyen [2], Tony Luczak [3], John E. Ball [2], Adam C. Knight [1], Brian K. Smith [3], Reuben F. Burch V [3] and R. K. Prabhu [4]

1. Department of Kinesiology, Mississippi State University, Mississippi State, MS 39762, USA; ems664@msstate.edu (E.S.); aknight@colled.msstate.edu (A.C.K.)
2. Electrical and Computer Engineering, Mississippi State University, Mississippi State, MS 39762, USA; dns105@msstate.edu (D.S.); ptn38@cavs.msstate.edu (P.N.); jeball@ece.msstate.edu (J.E.B.)
3. Department of Industrial Systems Engineering, Mississippi State University, Mississippi State, MS 39762, USA; luczak@ise.msstate.edu (T.L.); bks12@msstate.edu (B.K.S); burch@ise.msstate.edu (R.F.B.V.)
4. Department of Agricultural and Biomedical Engineering, Mississippi State University, Mississippi State, MS 39762, USA; rprabhu@abe.msstate.edu

* Correspondence: hchander@colled.msstate.edu; Tel.: +1-662-325-2963

Received: 19 August 2019; Accepted: 21 September 2019; Published: 24 September 2019

Abstract: Background: An induced loss of balance resulting from a postural perturbation has been reported as the primary source for postural instability leading to falls. Hence; early detection of postural instability with novel wearable sensor-based measures may aid in reducing falls and fall-related injuries. The purpose of the study was to validate the use of a stretchable soft robotic sensor (SRS) to detect ankle joint kinematics during both unexpected and expected slip and trip perturbations. Methods: Ten participants (age: 23.7 ± 3.13 years; height: 170.47 ± 8.21 cm; mass: 82.86 ± 23.4 kg) experienced a counterbalanced exposure of an unexpected slip, an unexpected trip, an expected slip, and an expected trip using treadmill perturbations. Ankle joint kinematics for dorsiflexion and plantarflexion were quantified using three-dimensional (3D) motion capture through changes in ankle joint range of motion and using the SRS through changes in capacitance when stretched due to ankle movements during the perturbations. Results: A greater R-squared and lower root mean square error in the linear regression model was observed in comparing ankle joint kinematics data from motion capture with stretch sensors. Conclusions: Results from the study demonstrated that 71.25% of the trials exhibited a minimal error of less than 4.0 degrees difference from the motion capture system and a greater than 0.60 R-squared value in the linear model; suggesting a moderate to high accuracy and minimal errors in comparing SRS to a motion capture system. Findings indicate that the stretch sensors could be a feasible option in detecting ankle joint kinematics during slips and trips.

Keywords: falls; slips; trips; postural perturbations; wearables; stretch-sensors; ankle kinematics

1. Introduction

Falls are one of the leading causes of both fatal and nonfatal injuries in clinical [1], geriatric [1], occupational [2], and healthy athletic populations [3] and can be induced due to environmental factors as well as physical and psychological human factors [4]. Subsequently, fall prevention methods and interventions are carried out by a wide range of healthcare and rehabilitation professionals [1]. With the advent of technology, multiple smart tools and applications have been used to combat falls using fall prevention intervention and specifically for detecting and diagnosing falls and fall risk [1]. Detecting falls or fall risks during everyday activities through activity monitoring can be beneficial during both prefall intervention, for individuals who are at fall risk, and postfall intervention, and for individuals who have already sustained a fall, in order to reduce their risk of subsequent falls [1]. While fall injury prevention sensor systems focus on responding to a fall after it has occurred and aid in contacting emergency medical assistance, fall detection sensor systems attempt to identify discrete fall events over the course of the day [1] and subsequently detect fall risk.

Wearable sensors have been used for human activity monitoring in various fields such as sports, training, fitness for improving performance, and preventing injuries and additionally, have also been used successfully in monitoring physical activity in clinical, pathological, and aging populations [5]. These wearable devices include different types of sensors such as inertial motion sensors (IMUs) [6,7], accelerometers [8,9], gyroscopes, magnetometers, switches, pedometers, goniometers, and foot pressure sensors that can provide kinematic and kinetic information of the body's movement [4,10–12]. Additionally, these devices have also been used in conjunction with other smart tools such as smart phones, smart shoes, modern camera systems, and even low-cost infrared thermal imaging sensors [13–17]. However, these sensors also have their own limitations, with a critical one being IMU distortion and drift [18,19], which can lead to an inaccurate representation of human activity monitoring. With the fast growth of sensor technology, several challenges towards design, development, fabrication, implementation, and utilization for continuous monitoring exist [9]. Recently, Luczak et al. [20] reported the current status of lack of wearbale solutions to accurately capture data "from the ground up" and the need for "closing the wearbale gap" through development and validation of novel types of sensors. Although this was specific to sensors used for the athletic population [20], there has also been a need for developing and validating novel sensors for fall detection, which is a leading cause for fatal and nonfatal injuries across different populations [1–4]. Hence, development and validation of other forms and types of wearable sensors to monitor human activity with more accuracy and less limitations, specifically for fall detection is required to close the wearable gap.

A hierarchy of approaches for fall detection has been previously proposed that includes camera-based systems to assess change in body shape, inactivity detection or three-dimensional (3D) head motion analysis, an ambience device that determines posture and presence, and wearable devices that evaluate posture and motion [5]. However, the camera-based systems and ambient device systems have their own limitations [6] such as capture obstruction, privacy concerns, false alarms, battery life, and sole intended use of the device [7,8]. Previous literature has reported that wearable devices can successfully detect induced falls in a laboratory setting [21] or other indoor environments [8]. Subsequently, different types of body-worn or wearable sensors appear to be the prominent choice for fall detection [5,10,22]. Early detection of fall risk, near falls, and incidences of falls classified by types (slip or trip induced) using wearable sensor technology can help aid in minimizing fall and fall-related injuries [11,21–23]. Due to higher precision, lower time commitment, easy administration, and feasibility, wearable biomechanical sensors are becoming popular for early detection of falls [8]. A recent review paper by Rucco et al. [22] addressed the impact of wearable sensors in fall detection by reporting the average number/age of participants; number of sensors, type of sensors, and their placement used in such fall detection studies. The predominant sample of populations tested included young and old individuals with age groups of less than 30 years of age and more than 64 years of age, and used a sample size of less than 10, 10–19, and 20–100 more commonly [22]. The most commonly used type of sensor being an accelerometer (more than 70%), followed by pressure sensors and gyroscopes, magnetometers with one or two sensors, predominantly placed and located on the trunk, foot, and leg [22]. More recently, a stretchable soft robotic sensor (SRS) that records a change in resistance values when stretched was used to determine if ankle joint-type movements could be inferred by using a custom-built rigid-body ankle joint mechanical device [20]. Based on the findings from this study, the SRS was capable of providing significant linear models in predicting sagittal plane ankle joint movement specific to plantarflexion [20]. As an extension of this research, a follow-up study by the same researchers successfully used similar stretch sensors that record capacitance change in response to stretch, to identify and detect ankle joint movements of plantarflexion (PF), dorsiflexion (DF), inversion (INV), and eversion (EVR) in human participants during non-weight-bearing isolated ankle movements [24]. Results from these research studies are published as "Closing the wearable gap: Part I and Part II" [20,24]. However, these stretch or flexible sensors have not yet been utilized to identify ankle joint movements in the more dynamic range, such as during slips and trips for fall detection.

Placement and position of wearable sensors used for quantifying body movements, balance, gait, and overall physical activity vary greatly across different parts of the body, ranging from the upper torso, lower torso, and lower extremities [10]. Specific to fall detection, the most commonly used sensor placement position includes the waist or hip, followed by trunk attachments [10]. Head and neck placements have also been used to assess acceleration patterns of the head during falls [25]. A higher success in detecting falls has been achieved by placing wearable sensors at the center of mass of the body [7]. However, the human body is considered as an inverted pendulum during upright balance maintenance, with the ankle joint serving as the axis of rotation [26]. Hence, placing wearable sensors on the foot and ankle segment can aid to capture recoveries and falls from a distal-to-proximal direction (ankle strategy) [26]. Previous research has used IMU sensors placed at the left and right ankle and sternum to successfully classify fall types based on slips and trips [27]. However, the use of an SRS sensor placed at the ankle and foot segment in detecting falls has not been analyzed.

Falls due to slips and trips are induced by a postural perturbation to the human body [28,29]. A postural perturbation is a sudden change in the orientation of the body that causes body disequilibrium and may lead to the displacement of the total body center of mass [26], thereby contributing to falls. One of the primary needs for fall detection is the assessment of postural responses during unexpected and expected postural perturbations, be it the "closed-loop" feedback postural control system when the external perturbations are unexpected and governed by an anticipatory sensory–motor, or "open-loop" feedforward postural control system when the external perturbations are expected [28]. Falls in the backward and forward directions are commonly studied in both real-world falls and simulated falls [10]. Simulated falls in a closed and controlled environment have been commonly used to analyze falls using fall prevention harness systems to protect the participants from any undesired falls. A systematic review on fall detection with body-worn sensors reported that 90 different studies (93.8% of the studies) used simulated falls [10]. Biomechanical analyses of human movement have evolved from simple goniometric measures to technologically advanced optical three-dimensional (3D) motion capture systems, with the latter seen as the gold-standard measure. However, the combination of SRS with ankle–foot placements during different postural perturbations in detecting falls, validated against a 3D motion capture system, has not been examined. Therefore, the purpose of the study was to validate the use of a stretchable SRS against a 3D motion capture system to identify ankle joint kinematics during both unexpected and expected slip and trip perturbations for fall detection. It was hypothesized that the SRS would be a valid tool for detecting ankle joint movements during both unexpected and expected postural slip and trip perturbations.

2. Materials and Methods

2.1. Participants

A total of 10 healthy adults (5 males and 5 females; age: 23.7 ± 3.13 years; height: 170.47 ± 8.21 cm; mass: 82.86 ± 23.4 kg) with no self-reported history of any musculoskeletal, neurological, cardiovascular, or vestibular disorders were recruited for the study. The participants' physical fitness status was also above recreationally trained (>3–4 days/week with consistent aerobic and anaerobic training for a minimum of the last 3 months leading up to testing). The average foot size of all participants was a size 10, and all males used the large/extra-large size socks, while all females used the small/medium size socks. Informed consent was obtained from the participants based on the approved protocol from the University's Institutional Review Board (IRB protocol # 18-121) after fully explaining the protocol along with the risks and benefits.

2.2. Instrumentation and Testing Environment

All experimental procedures were conducted in the University's Neuromechanics Laboratory. Biomechanical analysis of bilateral (left and right) ankle joint angular kinematics was assessed using a 3D motion capture system (Motion Analysis Corporation, Santa Rosa, CA, USA) using the Cortex software (Version 7.2.6). Four separate SRS (StretchSense, Auckland, New Zealand) were used to create an ankle fall sensor system (two stretch sensors on each leg) capable of detecting bilateral ankle joint movements that occur predominantly in the sagittal plane (ankle dorsiflexion and plantarflexion) [24]. The motion capture data were sampled at 100 Hz, and the SRS data were sampled at 25 Hz. An oscilloscope was used to measure the output from the StretchSense board. A sinusoidal voltage signal was applied for measuring the sensor capacitance. Characteristics of this signal include the following: Peak-to-peak voltage: 1.2 V; Frequency: 250 Hz; RMS: 417 mV; DC Offset: 1.6V was observed. A Burdick treadmill (Kone Instruments Inc., Milton, WI, USA) was used to provide the slip and trip perturbation, and the start and stop of the perturbation was controlled using the TA 520 treadmill controller (Kone Instruments Inc., Milton, WI, USA). All participants also wore a standardized safety harness (Protecta PRO harness) that meets or exceeds Occupational Safety and Health Administration (OSHA) 1910.66, OSHA 1926.502, American National Standard Institute (ANSI) A10.32, ANSI Z359.1, and ANSI Z359.3 during the slip and trip trials.

2.3. Experimental Procedures

All participants visited the Neuromechanics Laboratory for testing. The first 15–20 min was treated as the familiarization session, during which participants had their anthropometry measurements taken and were given an opportunity to get exposed to the fall safety harness and two trials each of slip and trip perturbations when standing on the treadmill (explained under procedures). The second session, following the familiarization was treated as the experimental testing session. Participants were provided with athletic compression garments, and reflective markers were placed on the lower extremity using a modified Helen Hayes model for the lower extremity [30]. Additionally, participants had four SRS, two each on the anterior and posterior side of each foot–ankle segment, placed in position using a compression sleeve. The SRS arrangement used in the current study is depicted in Figure 1.

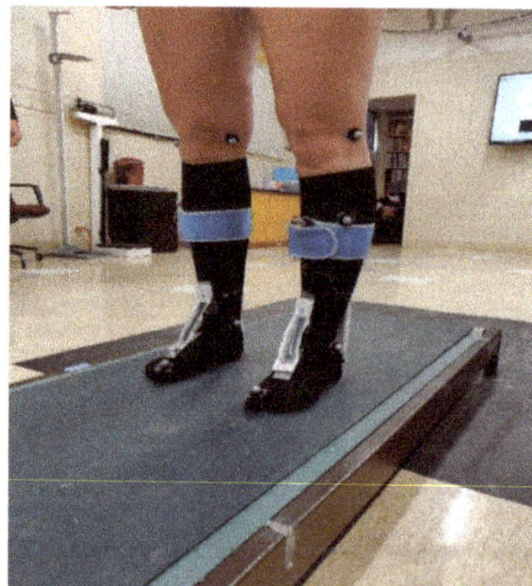

Figure 1. Arrangement of SRS and motion capture marker setup to capture and assess sagittal plane ankle kinematics (PF and DF), positioned on a treadmill belt to provide external postural perturbation in the forward and backward direction. SRS: soft robotic sensor. PF: plantarflexion. DF: dorsiflexion.

Participants were then directed to stand on the treadmill, and the fall-arrest harness system was attached to protect from undesired falls due to the treadmill perturbations. Participants were instructed to stand erect, being as still as possible each time before the beginning of slip or trip trials, during which the treadmill belt was turned on and off to provide a short brief postural perturbation. The treadmill was operated manually at a preset velocity of 0.67 m/s to provide both slip and trip perturbations. The participant was standing facing backward on the treadmill for slip perturbations (Figure 2) and standing facing forward on the treadmill for trip perturbations (Figure 3). All participants were exposed to one unexpected slip (US) and one unexpected trip (UT) in a counterbalanced order, followed by a series of three expected slips (ES) and a series of three expected trips (ET) in counterbalanced order, with randomized time intervals between the three slip and trip trials. During US and UT trials, participants were not aware of the time of perturbation, which was randomly provided within a 20 s interval to replicate an unexpected postural response through feedback postural control. The timing of the perturbation was random, and the instructions to the participants were always the same for US and ES, during which participants were instructed to stand as erect and still as possible. However, during ES and ET trails, participants were provided a countdown of three seconds before initiation of the treadmill perturbation, to replicate an expected postural response through feedforward postural control. Completion of two unexpected trials (one US and one UT) and six expected trials (three ES and three ET) marked the completion of experimental testing.

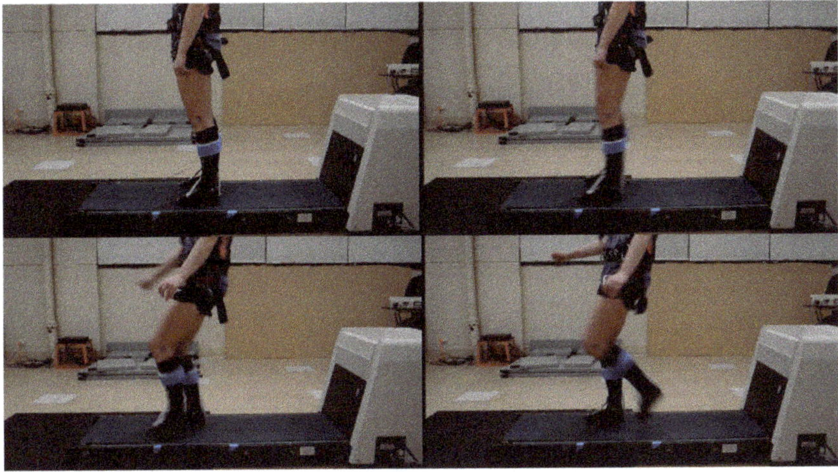

Figure 2. A sequence of an unexpected slip perturbation with wearable SRS and a motion capture marker system to assess fall detection.

2.4. Data Analysis and Statistical Analysis

Motion capture kinematic data for ankle joint range of motion was determined using a modified Helen Hayes model for the lower extremity through Cortex software. Raw kinematic data were filtered with a low-pass third-order Butterworth filter with a cut-off frequency of 30 Hz. The raw capacitance values of the SRS were measured using the 10 Channel Serial Peripheral Interface (SPI) Sensing Circuit in conjunction with the Bluetooth Low Energy (BLE) module, both made by StretchSense. The values were recorded using the proprietary StretchSense BLE iOS application. Due to the nature of the study, every trial produced a unique response for motion capture data and SRS data. In order to conduct the analysis of each trial in a consistent manner, data was only observed from the beginning of the slip or trip perturbation until the joint angle returned to its baseline value prior to the perturbation, which was identified as the "base angle". The absolute max joint angle that occurred in each trial was identified as the "peak angle." Similar peak and base values were collected for the SRS. Each trial consisted of joint angle data collected from the motion capture system and capacitance data collected from the StretchSense module. Peak values were noted for each trial, being either in DF or PF, depending on whether the joint angle increased or decreased in value upon activation of the treadmill perturbation. The difference between the peak value and base value was calculated to indicate the range of motion (ROM) and the capacitance change that occurred for each trial. For motion capture range of motion, negative values indicate PF ROM and positive values indicate DF ROM. Additionally, each trial dataset was scaled down and verified to ensure that the data was adjusted adequately over time and formatted for both motion capture and StretchSense data. For each trial, a model depicting SRS capacitance versus joint angle was created for each foot based on whether the foot initially went into PF or DF (for slip trials and trip trials, respectively). For PF, the posterior SRS was analyzed, while the anterior SRS was analyzed for DF.

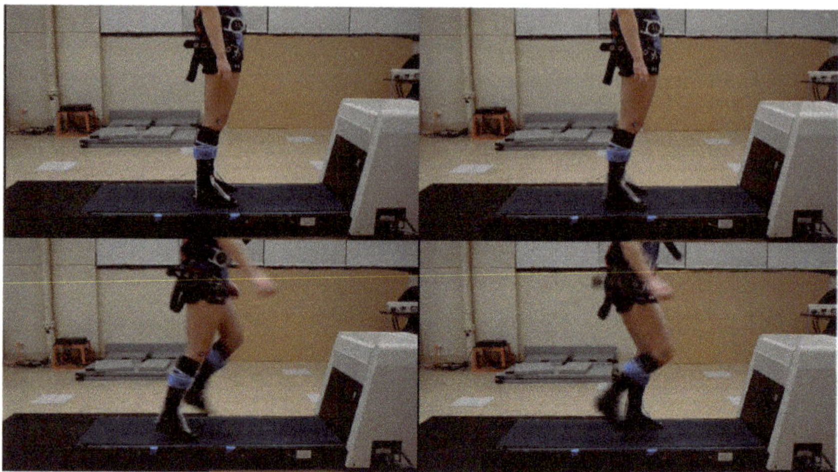

Figure 3. A sequence of an unexpected trip perturbation with wearable SRS and a motion capture marker system to assess fall detection.

An R (statistical computing software) script was used to generate the linear models and calculate adjusted R-squared and root mean square error (RMSE) values to determine a relative and absolute goodness of fit. A detailed description of the linear model comparing motion capture data and stretch sensor data is provided in Part I and Part II of "closing the wearable gap" papers, recently published from the same researchers in 2018 and 2019, respectively [20,24]. These measures provided metrics to indicate how well the SRS modeled the ankle joint movement during the slip and trip perturbations. For each trial, the base angle (i.e., the joint angle value immediately before perturbation occurred) and peak angle (most extreme angle value in first response) were analyzed for both feet. Additionally, the base and peak capacitance were analyzed for each SRS. The movement that occurred first for each foot (i.e., PF or DF) was noted for each trial. A difference was calculated between all the peak and base values, producing a total joint ROM that occurred during the trial as well as total capacitance change. An example of a bad processing trial that produced bad/poor results and an example of a good processing trial that produced good/great results are depicted in Figures 4 and 5, respectively. A bad processing trial rendered an R-squared value of 0.7524, and a good processing trial rendered an R-squared value of 0.9781. Finally, additional comparisons such as peak joint angle value comparisons across both feet as an average for each trial and comparisons between males and females were performed to see if there was an observed difference in range of motion and capacitance change and to identify how they affect range of motion as well SRS modelling performance. PF and DF movements were contrasted to see if one type of movement was easier to model with the SRS than the other movement.

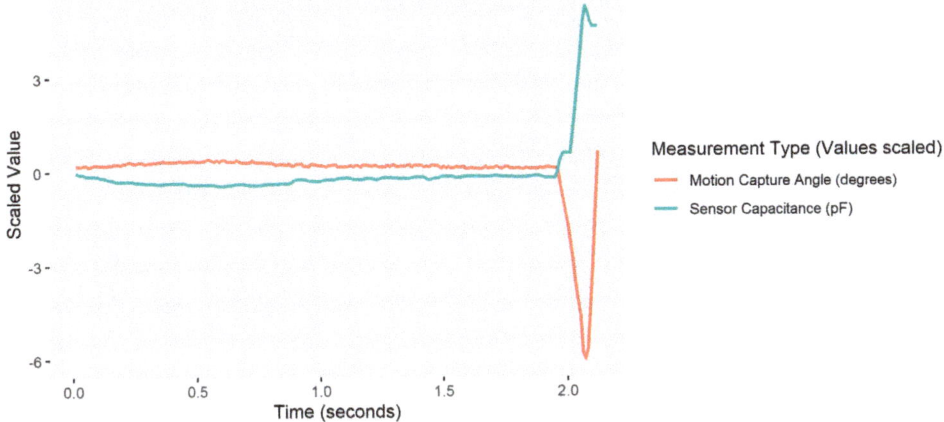

Figure 4. An example of a bad processing trial that produced bad/poor results.

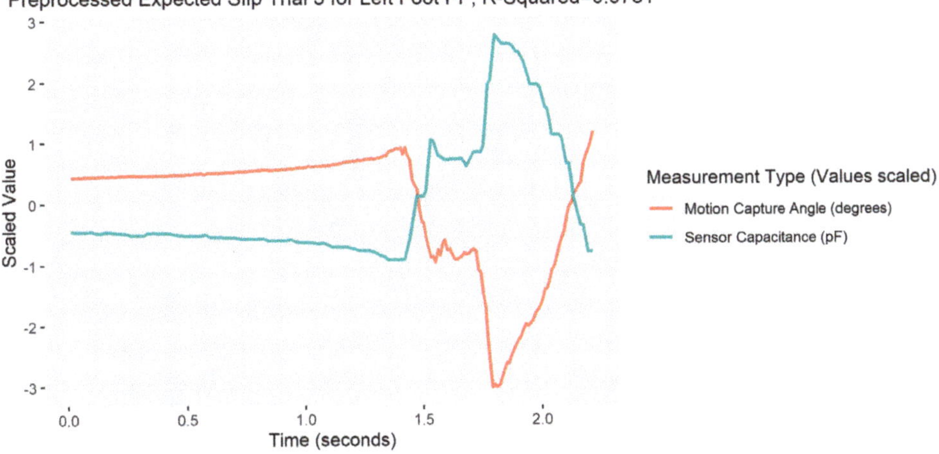

Figure 5. An example of a good processing trial that produced good/great result.

3. Results

Legends: US—Unexpected Slip, ES1—Expected Slip Trial 1, ES2—Expected Slip Trial 2, ES3—Expected Slip Trial 3, UT—Unexpected Trip, UT1—Unexpected Trip Trial 1, UT2—Unexpected Trip Trial 2, UT3—Unexpected Trip Trial 3.

Results from both the motion capture data and the SRS data were used to identify ankle joint ROM change from base angle to peak angle and change in capacitance from the base angle position to peak angle position. A series of violin plots are used to present the observed data from the slip and trip trials for all ten participants. The violin plots provide curved areas for each of the slip and trip trials that provides an idea of the "spread" of the data. These violin plots represent a kernel density distribution portrayed vertically. A greater horizontal width of a curve in the plot indicates a greater portion of participants that produced results near the value on the y-axis [24]. Additionally, the presented individual participant data points reiterate the "spread" of the data, so that outliers can be easily identified as well.

In Figures 6 and 7, the average and spread of ankle ROM and peak ROM in plantarflexion for slip trials and dorsiflexion for trip trials are presented respectively. Figures indicate the behavior of ankle joint movement going into plantar flexion during slip perturbations and going into dorsiflexion during trip perturbations. In Figure 8, the average and spread of capacitance change for each foot across every slip and trip trial are presented, to demonstrate the sensor output data.

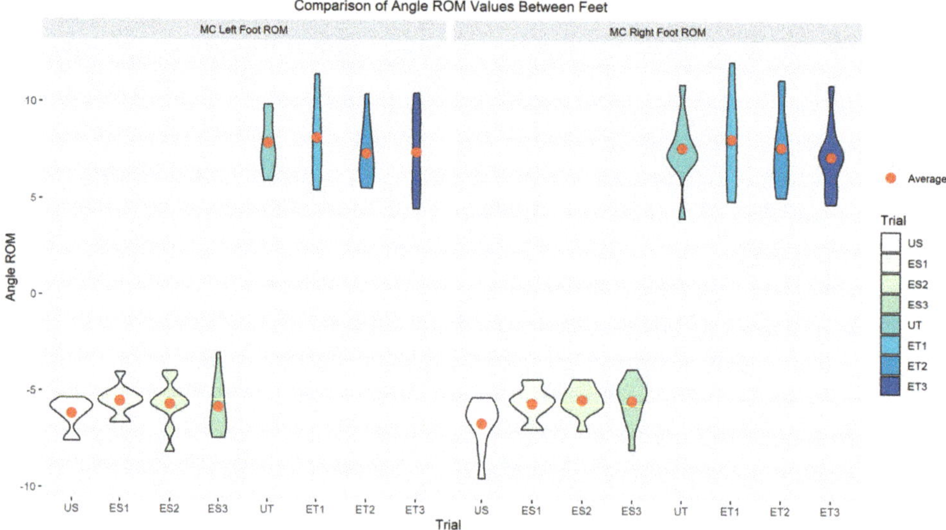

Figure 6. Spread and average ankle range of motion change for each foot across every trial; note that positive and negative changes occur respective to the trial with negative values indicating movement of ankle into PF and positive values indicate movement of ankle into DF.

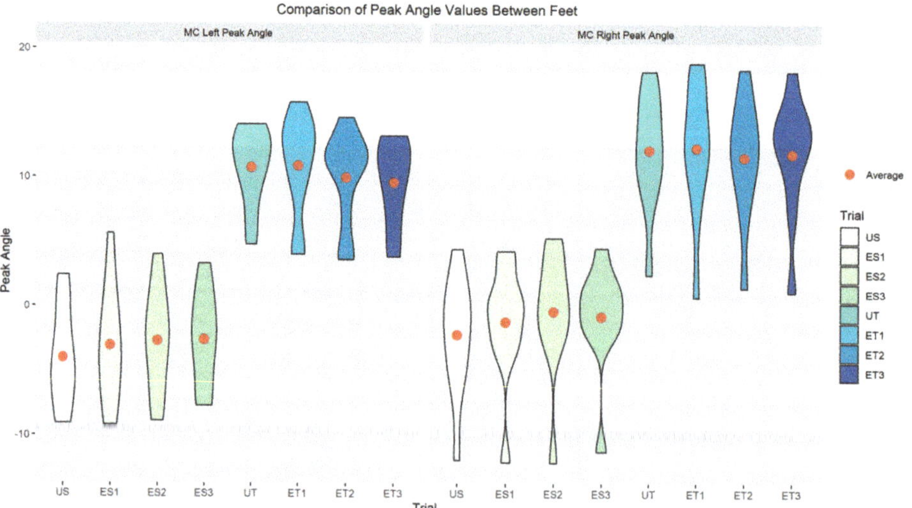

Figure 7. Spread and averages peak angles for each foot across every trial; note that positive and negative changes occur respective to the trial with negative values indicating movement of ankle into PF and positive values indicate movement of ankle into DF.

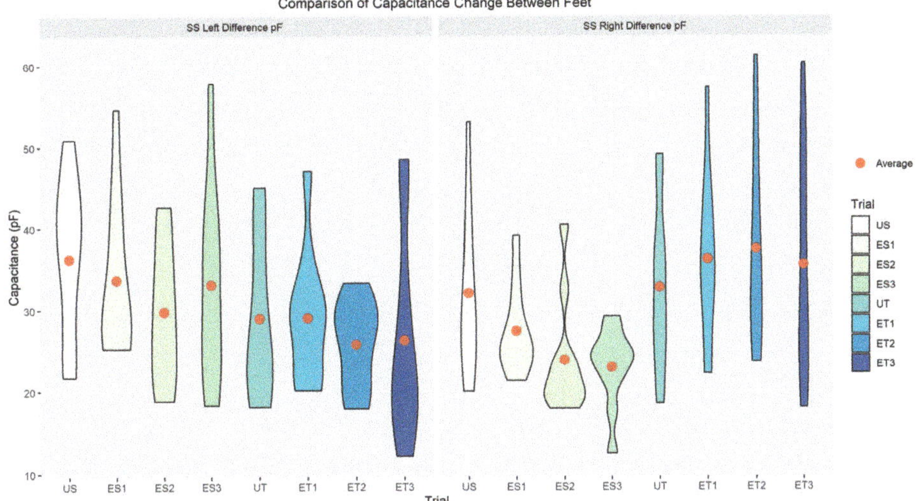

Figure 8. Spread and averages of capacitance change for each foot across every trial.

Results from the study are also presented in different pictorial representations indicating the performance of each trial based on adjusted R-squared and RMSE values (Figure 9; Figure 10). Average adjusted R-squared and average RMSE were identified for determining a relative and absolute goodness of fit of the model for the comparison of motion capture change in ankle angles versus the change in capacitance for all trials. On average, all trials had greater adjusted R-squared values and lower RMSE values in the linear model for the goodness of fit (Figures 9 and 10). Based on the violin plots in Figures 9 and 10, a greater portion of participants produced an R-squared value of more than 0.75 (moderate to high accuracy) and a greater portion of participants produced a RMSE value of lower than 4 (minimal errors). For the left foot–ankle kinematic detection, the highest adjusted R-squared value was 0.9781 (average = 0.7658) and the lowest RMSE was 1.0638 degrees (average = 3.1319 degrees). For the right foot–ankle kinematic detection, the highest adjusted R-squared value was 0.9832 (average = 0.7362) and the lowest RMSE was 0.8176 degrees (average = 0.9832 degrees). Results from the study demonstrated that 71.25% of the trials exhibited a minimal error of 4.0 degrees difference from the motion capture system and a greater than 0.60 R-squared value in the linear model, suggesting a moderate to high accuracy and minimal errors in comparing SRS with a motion capture system.

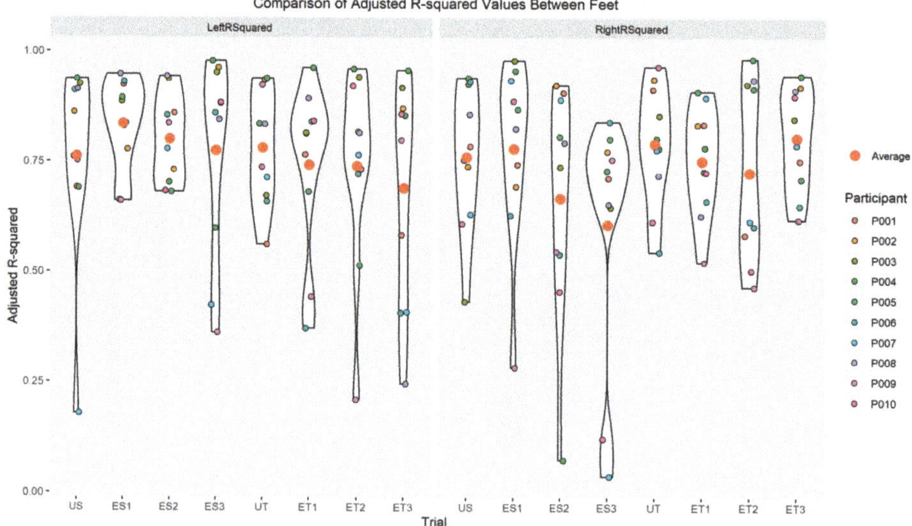

Figure 9. Spread, average, and individual data points of R-squared values for each trial across feet.

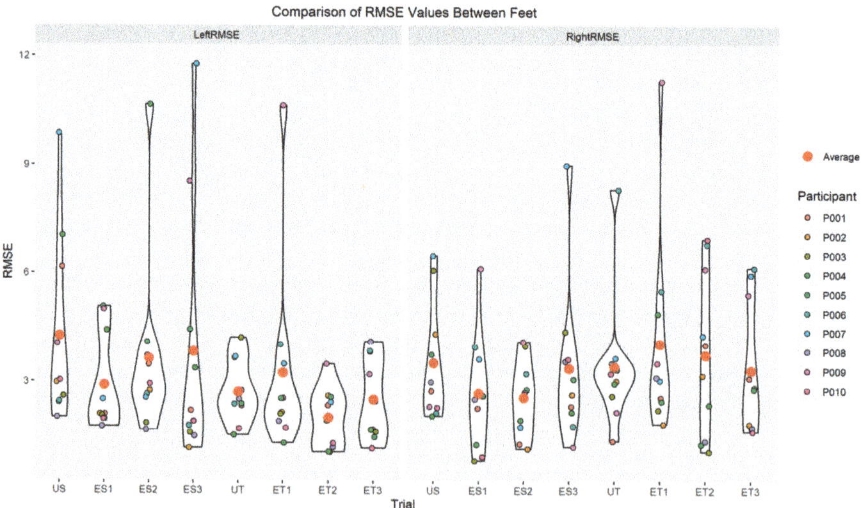

Figure 10. Spread, average, and individual data points of RMSE values for each trial across feet.

4. Discussion

The purpose of the study was to validate the use of SRS against a 3D motion capture system and to identify ankle joint kinematics during both unexpected and expected slip and trip perturbations for fall detection. SRS was hypothesized to be a valid tool for detecting ankle joint movements during both unexpected and expected postural slip and trip perturbations. Results from the current study indicated SRS as a viable product to detect ankle joint kinematics during unexpected and expected slips and trips and potentially serve as an early fall detection device. This was evident from the observed results, indicated by a high adjusted R-square value and low RMSE in the goodness of fit model between motion capture kinematics and SRS capacitance data during slip–trip trials.

During a slip perturbation, the ankle joint moves into a PF position as the center of mass (COM) of the human body is forced outside of the standing base of support (BOS) area in the posterior direction (leaning backward) [26,28,29]. During a trip perturbation, the ankle joint moves into DF position as the COM is forced outside the BOS in the anterior direction (leaning forward) [26,28,29]. Moreover, during such postural perturbations, not having the knowledge (unexpected/unanticipated) and having the knowledge (expected/anticipated) of the perturbation influences the biomechanics of fall recovery [31–34]. Unanticipated recoveries work using feedback postural control, whereas anticipated recoveries work using feedforward postural control. Results from the current study, as demonstrated in Figures 6 and 7, support the behavior of the ankle joint during slip and trip postural perturbations. Comparison of ankle ROM change and peak, quantified by a motion capture system and capacitance change, measured by stretch sensors during these unexpected and expected slips and trips supported previous fall research conducted through motion capture technology [35,36]. Additionally, the results from the current study support previous literature regarding the feasibility of using wearable sensors for fall detection [22]. However, the primary purpose of the current study was to assess if stretch sensors could be used to identify ankle joint kinematics during such slip–trip perturbations to detect falls.

The current findings from this slip–trip study, presented here as Part III, are an extension of Part I and Part II papers from two previous studies from the current researchers on "closing the wearable gap" projects, in an attempt to develop feasible but accurate sensors using stretchable SRS for human movement monitoring from the ankle joint and above (from the ground up) [20,24]. The use of SRS was reported to produce a linear model for PF movement using a custom-built ankle joint device from Part I of "Closing the wearable gap" [20]. Additionally, testing the SRS on human participants to detect ankle joint movements compared with motion capture data was successful and reported to accurately detect ankle joint PF, DF, INV, and EVR movements using four stretch sensors from Part II of "Closing the wearable gap" [24]. The primary aim in Part II of the paper [24] was to test the soft robotic sensors for placement and orientation on the foot and ankle segment. Because the foot and ankle segment is a complex human joint (capable of triaxial movements), the orientation and placement/location of the sensors was crucial to get the accurate measurements of the movements possible. Hence, a total of 10 positions/locations and orientations were compared to identify the most desirable location for accurate movements. Additionally, due to the complexity of the foot and ankle movements, the following were the testing conditions: only isolated movements (one at a time) of ankle dorsiflexion, plantarflexion, inversion, and eversion; only in non-weight-bearing conditions, meaning not making contact with the ground, and only in one side of the foot (right side). However, in the current study, the previously developed sensors are being used for an entirely new application with fall prevention. There have been multiple studies reporting the efficacy of using postural perturbation in studying falls for fall detection. There have also been multiple studies that have used wearable sensors for fall detection. However, to the author's knowledge, there has not been a previous study to validate the use of stretchable soft robotic sensors for fall detection (slips and trips). Additionally, the current project and paper addresses more real-life situations for falls that could be analyzed from a laboratory setting and is different from our Part II paper [24] in the following methods: combined movements of the foot and ankle movements with novel movement patterns of slipping and tripping; in weight-bearing conditions replicating slips and trips both without and with the knowledge of the individual; sensors on both sides of the feet to identify any asymmetries; validate the use of soft robotic sensors for fall detection that can be applied to all populations, ranging from geriatric to athletic and from clinical to occupational, all populations who are fall prone. Finally, the focus of this project and paper was to identify if these types of stretchable soft robotic sensors could be used for fall detection by measuring ankle range of motion, as these types of sensors have not been used for this purpose previously, at least to the author's knowledge.

Subsequently, using these wearable SRS for detecting falls and potentially creating a wearable fall detection device is much needed. The current study tested the use of SRS during simulated real-life-type falls using backward (slip) and forward (trip) perturbations, both without (unexpected) and with (expected) the knowledge of the perturbation. Results from the current study supported the findings from Part I and Part II of the previous studies [20,24]. Based on the current findings, the use of SRS was found to have greater R-squared value and lower RMSE in the linear regression model, suggesting greater goodness of fit in comparing motion capture ankle joint kinematics with capacitance change from the SRS. The violin plots in Figures 9 and 10 demonstrate that a greater portion of participants produced an R-squared value of more than 0.75 (moderate to high accuracy) and a greater portion of participants produced an RMSE value of lower than 4 (minimal errors). The higher R-squared values and low RMSE were also evident when comparing all unexpected and expected slip and trip trials and across both feet as well. Results from this study indicate that the stretch sensors could be used as a feasible option in detecting falls during slips and trips, even when they are unexpected or expected and across both left and right foot–ankle segments. Results from the study demonstrated that 71.25% of the trials exhibited a minimal error of less than 4.0 degrees difference from the motion capture system (lowest RMSE = 1.06 degrees and average RMSE = 3.13 degrees for the left foot and lowest RMSE = 0.81 and average RMSE = 3.33 degrees for the right foot) and a greater than 0.60 R-squared (highest R-squared value was 0.9781 and average R-squared = 0.7658 for the left foot and highest R-squared value was 0.9832 and average R-Squared = 0.7362 for the right foot) value in the linear model, suggesting a moderate to high accuracy and minimal errors in comparing SRS with a motion capture system. The R-squared values and RMSE were also evident when comparing all unexpected and expected slip and trip trials and across both feet as well, suggesting that SRS was a feasible option to detect bilateral ankle joint movements during slip–trip perturbations, using a total of four sensors.

While motion capture technology aids assessment of the joint ROM with gold-standard precision measures [18], it is still majorly confined within a laboratory setting, with limited implications to everyday tasks. Moreover, the financial cost and time consumed are also greater with the use of laboratory-based motion capture equipment. Therefore, there is a great demand for alternative solutions to precisely measure joint kinematics outside of a laboratory that have lower financial and time cost and can capture day-to-day, real-life scenarios. A wearable device that can measure changes in joint ROM and limit the negative aspects of motion capture while being precise appears as a promising solution [37]. The current study's results offer unique findings in validating the use of wearable stretch sensors that can detect ankle joint ROM while minimizing limitations that exist with motion capture and other wearable devices for fall detection.

4.1. Limitations

Limitations of the study's experimental procedures included the exposure of slip and trip perturbations that are not similar to the real-world situation, as a manual trigger treadmill was used to induce perturbation from a stationary standing position and all participants were harnessed in a fall-arrest system to prevent undue falls during testing. However, every attempt was made to ensure that participants did not know the occurrence of the unexpected perturbations by providing the perturbations in a randomized time point and providing the same instructions to stand as still and erect as possible to the participants. Additionally, participants were also allowed one familiarization session to get acquainted with the harness system to generate a real-life fall recovery response. Limitations of the study's data analysis were largely due to the low sampling rate of the StretchSense software, as indicated with examples in the results section of this paper, especially with an example each for a bad processing trial and a good processing trial. While the SRS was tested for ankle movements during slip and trip perturbations, the current study did not test ankle movements during other forms of activity such as walking, running, jumping, and so forth. Hence, the results observed in the study cannot be directly applied to other forms of human activity. The consumer acceptance level for using SRS for fall detection will depend on the population and the types of falls, and the results from the study should be used as preliminary findings. Finally, the material properties of the SRS have not yet been analyzed, which may further have more limitations. However, a previous study has reported "excellent linear trend with little noise" and no hysteresis [38]. Finally, the SRS device is still a prototype and would require further refining, ruggedization, testing (especially beyond a laboratory), and incorporating real-life falls and environments to test the effectiveness of the wearable SRS fall detection solution.

4.2. Future Work

Future work on fall detection devices can incorporate SRS for identifying joint kinematics. Moreover, wireless sensor networks, algorithms, and machine learning techniques have been used along with accelerometers and IMUs for fall detection [6,8–10,39] and in the future can also be implemented using SRS. However, adding electromyography (EMG) for fall detection in addition to joint kinematics detection can increase the accuracy of pre-impact fall detection, using both biomechanical and neuromuscular measures. The concept of pre-impact fall detection has been suggested earlier in attempts for early fall detection by using inertial sensors and fall-threshold-detecting algorithms [6,7,22,40]. Pre-impact fall detection research has been successful in detecting fall events at least 70 ms before the impact with the ground [7] and with an average lead time of 700 ms before the impact occurs, with no false alarms [40]. Using IMUs, pre-falls are usually detected due to abnormal or aberrant movement patterns of body segments that occur during falls but do not necessarily occur during regular activities of daily living [7,40]. More recently, a machine learning approach using EMG from the lower extremity has been successful in detecting pre-falls with a lead time of about 775 ms before the fall impact on the ground for forward, backward, and lateral falls [41]. However, as reported in Rucco et al. [22], the use of an accelerometer as a fall detection sensor has been more common due to its low cost and easy application compared with other sensor approaches such as EMG which require more complex sensor positioning, measurement, and analysis. Subsequently more research is warranted with more types of sensors to detect falls more precisely and efficiently. A combination of wearable stretch sensors, as discussed in this study, and EMG sensors with a machine learning approach can potentially be used for fall detection. The current research team is working on incorporating biomechanical and neuromuscular measures as a wearable solution for detecting falls. Finally, not much research has been conducted on the material properties of the SRS. Future work should also focus on testing the stress–strain properties and attempt to incorporate devices such as nanogenerators that can produce current with no requirements of external power supply that can be a safe and viable option for wearable applications.

5. Conclusions

With falls and fall-related injuries posing a significant threat to multiple populations, such as clinical, geriatric, athletic, occupational, and the healthy, accurate detection of ankle joint movements during postural perturbations using wearable solutions is crucial. Results from the study demonstrated that 71.25% of the trials exhibited a minimal error of less than 4.0 degrees difference from the motion capture system and a greater than 0.60 R-squared value in the linear model, suggesting a moderate to high accuracy and minimal errors in comparing SRS with a motion capture system. Findings indicate that the stretch sensors could be a feasible option in detecting ankle joint kinematics during slips and trips. Findings from this project will help in identifying wearable solutions in early detection of fall risk and prevent falls and fall-related injuries.

Author Contributions: Conceptualization: H.C.; methodology: H.C.; investigation: E.S. and D.S.; resources, R.F.B.V., J.E.B., B.K.S., A.C.K. and R.K.P.; data analysis: H.C., E.S., D.S., P.N. and T.L.; writing—original draft preparation: H.C.; writing—review and editing, R.F.B.V., J.E.B., B.K.S., A.C.K. and R.K.P.; visualization: D.S., P.N., T.L. and J.E.B.; supervision: H.C.; project administration: H.C.; funding acquisition: H.C. and E.S.

Funding: This project and publication was supported by Grant #2T42OH008436 from the National Institute of Occupational Safety and Health (NIOSH).

Acknowledgments: This project and publication was supported by Grant #2T42OH008436 from the National Institute of Occupational Safety and Health (NIOSH). Its contents are solely the responsibility of the authors and do not necessarily represent the official views of NIOSH. The research presented in this paper was partly funded by the National Science Foundation under NSF 18-511 - Partnerships for Innovation award number 1827652. The authors sincerely thank the anonymous reviewers for their valuable suggestions for improving the manuscript.

Conflicts of Interest: The authors declare no conflict of interest. The funders had no role in the design of the study; in the collection, analyses, or interpretation of data; in the writing of the manuscript, or in the decision to publish the results.

References

1. Hamm, J.; Money, A.G.; Atwal, A.; Paraskevopoulos, I. Fall prevention intervention technologies: A conceptual framework and survey of the state of the art. *J. Biomed. Inform.* **2016**, *59*, 319–345. [CrossRef] [PubMed]
2. Injuries, Illnesses, and Fatalities. Available online: https://www.bls.gov/iif/ (accessed on 9 August 2019).
3. Chander, H.; Dabbs, N.C. Balance Performance and Training Among Female Athletes. *Strength Cond. J.* **2016**, *38*, 8. [CrossRef]
4. Chaccour, K.; Darazi, R.; Hassani, A.H.E.; Andrès, E. From Fall Detection to Fall Prevention: A Generic Classification of Fall-Related Systems. *IEEE Sens. J.* **2017**, *17*, 812–822. [CrossRef]
5. Mukhopadhyay, S.C. Wearable Sensors for Human Activity Monitoring: A Review. *IEEE Sens. J.* **2015**, *15*, 1321–1330. [CrossRef]
6. Genovese, V.; Mannini, A.; Guaitolini, M.; Sabatini, A.M. Wearable Inertial Sensing for ICT Management of Fall Detection, Fall Prevention, and Assessment in Elderly. *Technologies* **2018**, *6*, 91. [CrossRef]
7. Wu, G.; Xue, S. Portable Preimpact Fall Detector With Inertial Sensors. *IEEE Trans. Neural Syst. Rehabil. Eng.* **2008**, *16*, 178–183. [PubMed]
8. Chen, J.; Kwong, K.; Chang, D.; Luk, J.; Bajcsy, R. Wearable Sensors for Reliable Fall Detection. In Proceedings of the 2005 IEEE Engineering in Medicine and Biology 27th Annual Conference, Shanghai, China, 1–4 September 2005; pp. 3551–3554.
9. Srinivasan, S.; Han, J.; Lal, D.; Gacic, A. Towards automatic detection of falls using wireless sensors. In Proceedings of the 2007 29th Annual International Conference of the IEEE Engineering in Medicine and Biology Society, Lyon, France, 22–26 August 2007; pp. 1379–1382.
10. Schwickert, L.; Becker, C.; Lindemann, U.; Maréchal, C.; Bourke, A.; Chiari, L.; Helbostad, J.L.; Zijlstra, W.; Aminian, K.; Todd, C.; et al. Fall detection with body-worn sensors. *Z. Für Gerontol. Geriatr.* **2013**, *46*, 706–719. [CrossRef] [PubMed]
11. Ma, C.Z.-H.; Wong, D.W.-C.; Lam, W.K.; Wan, A.H.-P.; Lee, W.C.-C. Balance Improvement Effects of Biofeedback Systems with State-of-the-Art Wearable Sensors: A Systematic Review. *Sensors* **2016**, *16*, 434. [CrossRef] [PubMed]
12. Shany, T.; Redmond, S.J.; Narayanan, M.R.; Lovell, N.H. Sensors-Based Wearable Systems for Monitoring of Human Movement and Falls. *IEEE Sens. J.* **2012**, *12*, 658–670. [CrossRef]
13. Habib, M.A.; Mohktar, M.S.; Kamaruzzaman, S.B.; Lim, K.S.; Pin, T.M.; Ibrahim, F. Smartphone-Based Solutions for Fall Detection and Prevention: Challenges and Open Issues. *Sensors* **2014**, *14*, 7181–7208. [CrossRef] [PubMed]
14. Delahoz, Y.S.; Labrador, M.A. Survey on Fall Detection and Fall Prevention Using Wearable and External Sensors. *Sensors* **2014**, *14*, 19806–19842. [CrossRef] [PubMed]
15. Dai, J.; Bai, X.; Yang, Z.; Shen, Z.; Xuan, D. Mobile Phone-based Pervasive Fall Detection. *Pers. Ubiquitous Comput* **2010**, *14*, 633–643. [CrossRef]
16. Sixsmith, A.; Johnson, N. A smart sensor to detect the falls of the elderly. *IEEE Pervasive Comput.* **2004**, *3*, 42–47. [CrossRef]
17. Mellone, S.; Tacconi, C.; Schwickert, L.; Klenk, J.; Becker, C.; Chiari, L. Smartphone-based solutions for fall detection and prevention: the FARSEEING approach. *Z. Für Gerontol. Geriatr.* **2012**, *45*, 722–727. [CrossRef] [PubMed]
18. Fong, D.T.-P.; Chan, Y.-Y. The Use of Wearable Inertial Motion Sensors in Human Lower Limb Biomechanics Studies: A Systematic Review. *Sensors* **2010**, *10*, 11556–11565. [CrossRef] [PubMed]
19. Cooper, G.; Sheret, I.; McMillian, L.; Siliverdis, K.; Sha, N.; Hodgins, D.; Kenney, L.; Howard, D. Inertial sensor-based knee flexion/extension angle estimation. *J. Biomech.* **2009**, *42*, 2678–2685. [CrossRef] [PubMed]
20. Luczak, T.; Saucier, D.; Burch, V.R.F.; Ball, J.E.; Chander, H.; Knight, A.; Wei, P.; Iftekhar, T. Closing the Wearable Gap: Mobile Systems for Kinematic Signal Monitoring of the Foot and Ankle. *Electronics* **2018**, *7*, 117. [CrossRef]

21. Pang, I.; Okubo, Y.; Sturnieks, D.; Lord, S.R.; Brodie, M.A. Detection of Near Falls Using Wearable Devices: A Systematic Review. *J. Geriatr. Phys. Ther.* **2019**, *42*, 48. [CrossRef] [PubMed]
22. Rucco, R.; Sorriso, A.; Liparoti, M.; Ferraioli, G.; Sorrentino, P.; Ambrosanio, M.; Baselice, F. Type and Location of Wearable Sensors for Monitoring Falls during Static and Dynamic Tasks in Healthy Elderly: A Review. *Sensors* **2018**, *18*, 1613. [CrossRef] [PubMed]
23. Gordt, K.; Gerhardy, T.; Najafi, B.; Schwenk, M. Effects of Wearable Sensor-Based Balance and Gait Training on Balance, Gait, and Functional Performance in Healthy and Patient Populations: A Systematic Review and Meta-Analysis of Randomized Controlled Trials. *Gerontology* **2018**, *64*, 74–89. [CrossRef]
24. Saucier, D.; Luczak, T.; Nguyen, P.; Davarzani, S.; Peranich, P.; Ball, J.E.; Burch, R.F.; Smith, B.K.; Chander, H.; Knight, A.; et al. Closing the Wearable Gap—Part II: Sensor Orientation and Placement for Foot and Ankle Joint Kinematic Measurements. *Sensors* **2019**, *19*, 3509. [CrossRef] [PubMed]
25. Lindemann, U.; Hock, A.; Stuber, M.; Keck, W.; Becker, C. Evaluation of a fall detector based on accelerometers: A pilot study. *Med. Biol. Eng. Comput.* **2005**, *43*, 548–551. [CrossRef] [PubMed]
26. Winter, D. Human balance and posture control during standing and walking. *Gait Posture* **1995**, *3*, 193–214. [CrossRef]
27. Aziz, O.; Park, E.J.; Mori, G.; Robinovitch, S.N. Distinguishing the causes of falls in humans using an array of wearable tri-axial accelerometers. *Gait Posture* **2014**, *39*, 506–512. [CrossRef] [PubMed]
28. Horak, F.B. Clinical Measurement of Postural Control in Adults. *Phys. Ther.* **1987**, *67*, 1881–1885. [CrossRef] [PubMed]
29. Horak, F.B. Postural orientation and equilibrium: What do we need to know about neural control of balance to prevent falls? *Age Ageing* **2006**, *35*, ii7–ii11. [CrossRef] [PubMed]
30. Gait Analysis Models. Available online: http://www.clinicalgaitanalysis.com/faq/sets/ (accessed on 19 August 2019).
31. Chander, H.; Garner, J.C.; Wade, C. Heel contact dynamics in alternative footwear during slip events. *Int. J. Ind. Ergon.* **2015**, *48*, 158–166. [CrossRef]
32. Chander, H.; Garner, J.C.; Wade, C. Slip outcomes in firefighters: A comparison of rubber and leather boots. *Occup. Ergon.* **2016**, *13*, 67–77. [CrossRef]
33. Chander, H.; Wade, C.; Garner, J.C.; Knight, A.C. Slip initiation in alternative and slip-resistant footwear. *Int. J. Occup. Saf. Ergon.* **2017**, *23*, 558–569. [CrossRef]
34. Chander, H.; Knight, A.C.; Garner, J.C.; Wade, C.; Carruth, D.W.; DeBusk, H.; Hill, C.M. Impact of military type footwear and workload on heel contact dynamics during slip events. *Int. J. Ind. Ergon.* **2018**, *66*, 18–25. [CrossRef]
35. Chang, W.-R.; Leclercq, S.; Lockhart, T.E.; Haslam, R. State of science: Occupational slips, trips and falls on the same level. *Ergonomics* **2016**, *59*, 861–883. [CrossRef] [PubMed]
36. Shiratori, T.; Coley, B.; Cham, R.; Hodgins, J.K. Simulating balance recovery responses to trips based on biomechanical principles. In Proceedings of the 2009 ACM SIGGRAPH/Eurographics Symposium on Computer Animation-SCA '09, New Orleans, LA, USA, 1–2 August 2009; ACM Press: New Orleans, LA, USA, 2009; p. 37.
37. Chan, M.; Estève, D.; Fourniols, J.-Y.; Escriba, C.; Campo, E. Smart wearable systems: Current status and future challenges. *Artif. Intell. Med.* **2012**, *56*, 137–156. [CrossRef] [PubMed]
38. Litteken, D. Evaluation of Strain Measurement Devices for Inflatable Structures. In Proceedings of the 58th AIAA/ASCE/AHS/ASC Structures, Structural Dynamics, and Materials Conference, Grapevine, TX, USA, 9–13 January 2017; American Institute of Aeronautics and Astronautics: Grapevine, TX, USA, 2017.
39. Hakim, A.; Huq, M.S.; Shanta, S.; Ibrahim, B.S.K.K. Smartphone Based Data Mining for Fall Detection: Analysis and Design. *Procedia Comput. Sci.* **2017**, *105*, 46–51. [CrossRef]
40. Nyan, M.N.; Tay, F.E.H.; Murugasu, E. A wearable system for pre-impact fall detection. *J. Biomech.* **2008**, *41*, [illegible]. [CrossRef] [PubMed]
41. Rescio, G.; Leone, A.; Siciliano, P. Supervised machine learning scheme for electromyography-based pre-fall detection system. *Expert Syst. Appl.* **2018**, *100*, 95–105. [CrossRef]

Article

Recognition of Gait Phases with a Single Knee Electrogoniometer: A Deep Learning Approach

Francesco Di Nardo *, Christian Morbidoni *, Alessandro Cucchiarelli and Sandro Fioretti

Department of Information Engineering, Università Politecnica delle Marche, 60100 Ancona, Italy; a.cucchiarelli@univpm.it (A.C.); s.fioretti@staff.univpm.it (S.F.)

* Correspondence: f.dinardo@staff.univpm.it (F.D.N.); c.morbidoni@univpm.it (C.M.); Tel.: +39-071-220-4838 (F.D.N.); +39-071-220-4830 (C.M.)

Received: 27 January 2020; Accepted: 18 February 2020; Published: 20 February 2020

Abstract: Artificial neural networks were satisfactorily implemented for assessing gait events from different walking data. This study aims to propose a novel approach for recognizing gait phases and events, based on deep-learning analysis of only sagittal knee-joint angle measured by a single electrogoniometer per leg. Promising classification/prediction performances have been previously achieved by surface-EMG studies; thus, a further aim is to test if adding electrogoniometer data could improve classification performances of state-of-the-art methods. Gait data are measured in about 10,000 strides from 23 healthy adults, during ground walking. A multi-layer perceptron model is implemented, composed of three hidden layers and a one-dimensional output. Classification/prediction accuracy is tested vs. ground truth represented by foot–floor-contact signals, through samples acquired from subjects not seen during training phase. Average classification-accuracy of 90.6 ± 2.9% and mean absolute value (MAE) of 29.4 ± 13.7 and 99.5 ± 28.9 ms in assessing heel-strike and toe-off timing are achieved in unseen subjects. Improvement of classification-accuracy (four points) and reduction of MAE (at least 35%) are achieved when knee-angle data are used to enhance sEMG-data prediction. Comparison of the two approaches shows as the reduction of set-up complexity implies a worsening of mainly toe-off prediction. Thus, the present electrogoniometer approach is particularly suitable for the classification tasks where only heel-strike event is involved, such as stride recognition, stride-time computation, and identification of toe walking.

Keywords: knee angle; deep learning; neural networks; gait-phase classification; electrogoniometer; EMG sensors; walking; gait-event detection

---

1. Introduction

Modifications of motor function associated to different environments or state of health are typically estimated and quantified by means of instrumental gait analysis. To this aim, particularly relevant seems to be the problem of recognizing at least the two main gait phases, namely stance and swing. Single-type sensors or a combination of multiple types of sensors, such as angular velocity, attitude, force, electromyography, and cameras are typically used for gait phase quantification [1–5]. Recent availability of technological advancements is allowing to limit the experimental complexity of gait-analysis set-up, providing a less expensive, less intrusive, and more comfortable estimation of gait data. Robust artificial intelligence techniques for managing a lot of biological data and signals coming from smart sensors such as inertial measurements units (IMU) are undoubtedly among the most used approaches to this aim [6–14]. Specifically, the problem of estimating temporal parameters of gait could take great advantage by the development of these new approaches. Frequently, the use of IMUs appears to be suitable for a smart assessment of walking parameters, such as gait-phase duration and timing of heel strike (time when the foot touches the ground) and toe off (time when the foot-toes clear the ground) [11]. Attempts based on artificial intelligence were also applied in a satisfactory way for the assessment of gait parameters during walking [6,7,9,10,12–15].

Machine/deep learning techniques are usually implemented for classification of biological signals [13,16]. The introduction of those methodologies has opened a novel perspective also for reducing the complexity of experimental set-up. Predicting gait events from sensors already used in smart gait protocols or in specific environments would avoid the addition of further sensors or systems for the direct measurement of temporal data, such as stereo-photogrammetry, foot-switch sensors, pressure mats, and IMUs [11,17–19]. This would be particularly suitable for specific fields, such as walking-aid devices (mainly exoskeletons) where sensors are already embedded in the system [10,20–23]. From this point of view, an interesting attempt has been performed by Liu et al. [10], who proposed a technique for the recognition of gait phases using only joint angular sensors of the exoskeleton robot, containing the position, velocity, acceleration, and further motion data; very promising results were achieved. In the same way, different studies attempted to provide a reliable classification of gait phases and an accurate prediction of heel strike (HS) and toe off (TO) from only surface electromyographic (sEMG) sensors [13,14,21,23–25]. Details about methodology and outcomes of these studies are reported in Section 2 (related works).

Following the line taken by the above-mentioned studies, the present work was designed to propose a novel approach for the binary classification of gait phases and the prediction of gait events, based only on deep-learning analysis of sagittal knee-joint angle measured by a single electrogoniometer. Although promising classification performances were achieved by sEMG-based methods [13,14,21,23–25], gold-standard approaches are not available in literature. Thus, in order to evaluate the robustness of the proposed approach, a direct comparison in the same population was also performed with the sEMG-based experiment [14], which achieves the best performance in HS and TO prediction among the approaches reported in literature (see Section 2). Moreover, another aim of the study is to test if the addition of electrogoniometer data could further improve the classification performance of this state-of-the-art method.

2. Related Works

The gold standard in gait segmentation is nowadays represented by foot pressure insoles or by footswitches [17,26–28], which allow a direct measurement of foot–floor contact. Otherwise, IMUs and EMG signals are employed as input to gait-phase identification algorithms [11,13,14,21,23]. Recently, data fusion of sensors is suggested as a further reliable approach [29,30]. Artificial intelligence techniques are also satisfactorily employed for the estimation of walking parameters [6,7,9,10,12–15]. To the best of our knowledge, no studies attempting to classify/predict gait events from only sagittal knee angles are reported in literature. For the purposes of the present work, sEMG signals are of particular interest, being measured in every gait protocol in order to characterize the neuro-muscular activity and neuro-motor disabilities and being acquired very often together with kinematic data, such as sagittal knee angles.

Not so many efforts are available in literature, providing classification of gait phases from only sEMG signals [13,14,21,23–25]. Most of these studies aim only at classifying gait phases, not providing estimation of gait events (HS and TO). Joshi et al. introduced a control system for a foot-knee exoskeleton based on hand-crafted features computed from eight EMG signals to feed the Bayesian information criteria (BIC) [21]. Linear discriminant analysis (LDA) was then implemented to extract eight gait phases. One single subject was recruited for this experiment. The achieved accuracy ranged from 50% to 80%, with the combination of the BIC and LDA stage. Ziegier et al. employed a support-vector-machine classifier to provide binary segmentation of gait phases, based on a new bilateral feature (weighted signal difference) from EMG signal acquired in seven muscle pairs [23]. Only two subjects were used to test the approach, walking on a treadmill at different speeds. The accuracy ranged from 81% to 96% (mean value around 91%); maximum classification accuracy was identified when training and testing sets were strides from the same subject (intra-subject accuracy). Meng et al. used a hidden Markov model and set of EMG-based features to identify five gait sub-phases during treadmill walking [24]. Even in this study one single subject was used to test the classification. The best-case accuracy was 91.1%. The present group of researchers was able to achieve a mean binary-classification accuracy of 95.2%, adopting a multi-layer perceptron (MLP) classifier to interpret EMG data [25]. To this aim, an intra-subject approach was used on twelve healthy volunteers.

As far as we know, only two papers reported outcomes not only on classification of stance and swing but also on identification of heel strike and toe-off timing from sEMG signals [13,14]. Both studies adopted an inter-subject approach, consisting in training neural networks with sEMG signals measured during different strides of a population of homogeneous subjects and then testing the classifier on brand new subjects. Nazmi at al. extracted time-frequency EMG-based features to feed a single hidden layer neural network [13]. Training set was composed of seven subjects walking on a treadmill and testing set included one single unlearned subject. Mean classification accuracy of 87.5% for learned subjects and 77% for unlearned ones were accomplished. Prediction outcomes, computed in unseen subjects, achieved a mean absolute error (MAE ± SD) of 35 ± 25 ms in assessing HS and 49 ± 15 ms in assessing TO. The present group of researchers faced the same assignment, trying to interpret the linear envelopes extracted from sEMG signals by means of a multi-layer-perceptron classifier [14]. The MLP network was trained with sEMG data acquired during walking of 22 subjects and then tested on a brand-new subject. The procedure was performed twenty-three times, each time using a different subject as test-set (23-fold cross-validation). This approach provided an average (over 23-fold) binary classification accuracy of 94.9% for learned subjects and 93.4% for unlearned ones. *MAEs* in the prediction of HS and TO were 21 ± 7 ms and 38 ± 15 ms, respectively. This latter approach [14] is adopted as a reference experiment for the present study since it achieved the best performance in phase classification and gait event prediction among the inter-subject approaches reported in literature.

3. Materials and Methods

3.1. Participants

Foot–floor-contact, knee-angle, and surface EMG signals were recorded from 23 healthy adults (11 males and 12 females). Average volunteer characteristics (±SD) were: height = 173 (±10) cm; mass = 63.3 (±12.4) kg; and age = 23.8 (±1.9) years. Subjects have never presented pathological condition or undergone orthopedic surgery that might have affected leg mechanics. Moreover, volunteers with joint pain, neurological pathologies, and abnormal gait were not recruited. Overweight and obese subjects (body mass index ≥ 25) were excluded from the study. The present research was undertaken following the ethical principles of the Helsinki Declaration and was approved by local ethical committee.

3.2. Signal Acquisition

Signal acquisition was achieved by means of the multichannel recording system Step32 (Medical Technology, Italy, Version PCI-32 ch2.0.1. DV; resolution: 12 bit; sampling rate: 2 kHz). Volunteers were instrumented with one knee electrogoniometer, three foot-switches, and four sEMG probes for each leg. Experimental set-up is depicted in Figure 1. Then, they walked for around 5 minutes with bare feet at self-selected pace following an eight-shaped path, which includes natural deceleration, reversing, curve, and acceleration. Experiments were performed in the Motion Analysis Laboratory of Università Politecnica delle Marche, Ancona, Italy. An electro-goniometer (accuracy: 0.5°) was applied to the lateral side of each leg for measuring knee-joint angle in the sagittal plane

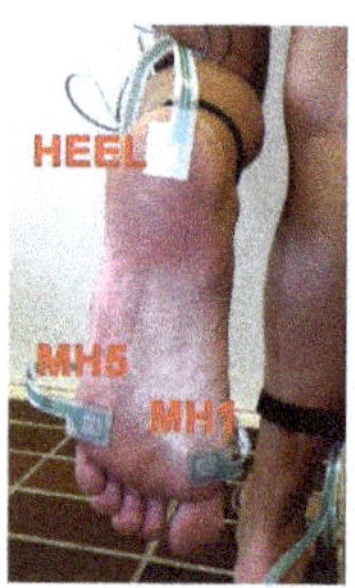

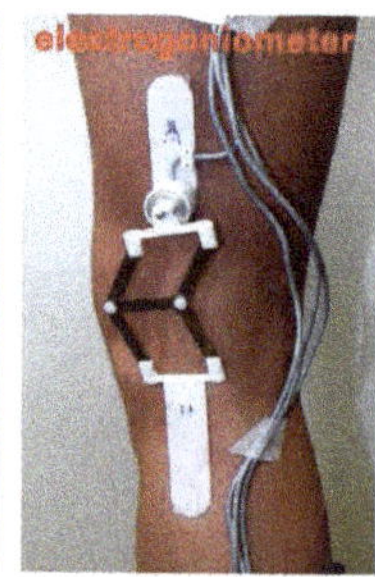
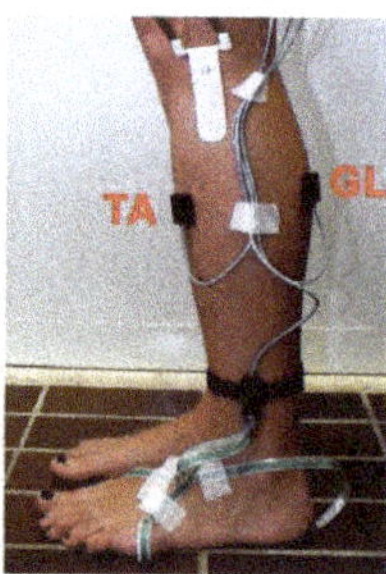

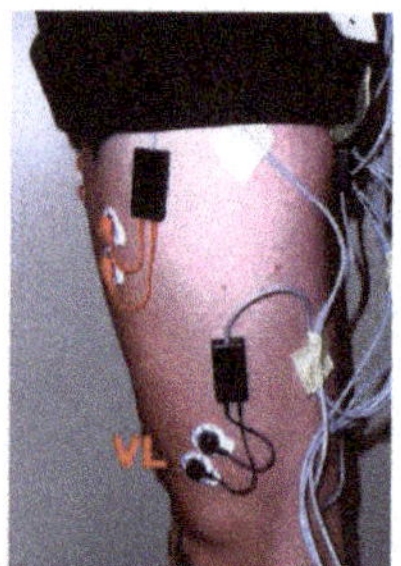

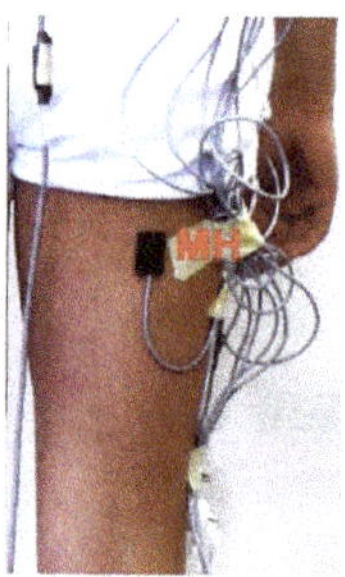

Figure 1. Experimental set-up. MH1 and MH5 are first and the fifth metatarsal heads, respectively. TA, tibialis anterior. GL, gastrocnemius lateralis. VL, vastus lateralis. MH, medial hamstrings.

Foot-switches were applied under the heel, the first and the fifth metatarsal heads of each foot for measuring foot–floor-contact signal. sEMG signals were registered by means of three single-differential probes with fixed geometry placed over tibialis anterior (TA), gastrocnemius lateralis (GL), and medial hamstrings (MH) and one further single-differential probe (minimum inter-electrode distance: 12 mm) with variable geometry placed over vastus lateralis (VL). Electrode location and orientation were carried out under the supervision of a skilled licensed physical therapist, complying with SENIAM recommendations [31]. sEMG signals are exactly the same collected for the previous study of the present group of researchers [14]. Characteristics of foot-switches are: size = 11 × 11 × 0.5 mm and activation force = 3 N. Foot-switch signals are used to identify stance/swing phases and HS and TO events considered as ground-truth.

3.3. Signal Pre-Processing

Knee angles in the sagittal plane measured by electrogoniometers were low pass filtered with cut-off frequency of 15 Hz. Each signal was min-max normalized within each subject, thus mapping the values in the [0–1] interval. An example of normalized flexion-extension knee angle in a representative subject is reported in Figure 2. Footswitch signals were processed for identifying the different gait cycles and phases (stance and swing), according to the approach discussed in [26]. sEMG signals were amplified, high-pass filtered (linear-phase FIR filter, cut-off frequency: 20 Hz) and low-pass filtered (linear-phase FIR filter, cut-off frequency: 450 Hz) for removing motion artefacts and high frequency noise, respectively. After a full-wave rectification, a second-order Butterworth low-pass filter (cut-off frequency: 5 Hz) was applied to extract the envelope of the signal, following the classic indication provided by acknowledged studies by Hermens et al. and Winter et al. [31,32]. Winter proposed a cut-off frequency of 3 Hz, while Hermens suggested a cut-off frequency of 10 Hz. The cut-off frequency of 5 Hz adopted in the present paper seems to be a good compromise between the two approaches. Finally, each sEMG signal was min-max normalized within each subject and for each muscle.

3.4. Data Preparation

Classification performances were tested after two different approaches for feeding the classifier: giving only knee-angle signal (knee approach) or giving knee-angle and sEMG signals (KEMG approach) as input to train the classifier.

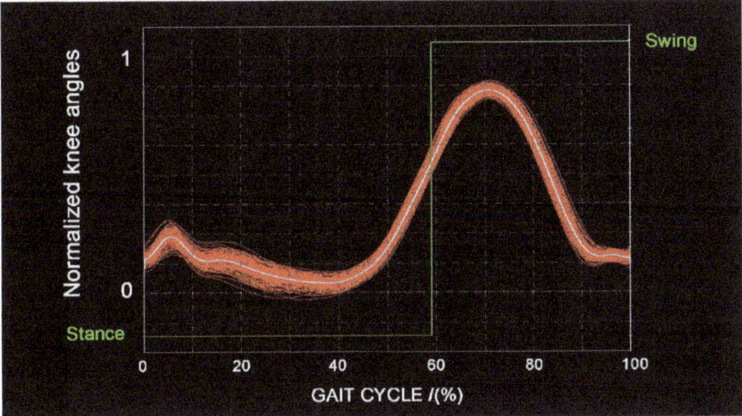

Figure 2. Example of normalized flexion-extension knee angle in sagittal plane measured in all the strides walked by a representative subject (red curves). Mean curve is depicted in white. Stance and swing durations are portrayed in green.

3.4.1. Knee Approach

For suitably training the classifier, each knee-angle signal was split into 20-sample windows (corresponding to 10 ms). A chronological sequence of 40-sample vectors was created, where each vector included the two synchronized 20-sample windows from two knee-angle signals (right and left leg). In details, the first sample of the first 40-sample vector of the sequence was the first sample of the knee angle measured in the right leg; the second sample of the first 40-sample vector was the first sample of the knee angle measured in the left leg.

Then, a specific label was assigned to each 40-sample window as follows: if the value of all the samples of the basographic signal corresponding to the 40-sample vector was 0 (or 1), a global label 0 (or 1) was assigned to the 40-sample vector. 40-sample vectors, including swing-to-stance or stance-to-swing transitions, were discarded. This approach including only knee-angle data to feed the neural network is referred to as 'Knee approach'.

3.4.2. KEMG Approach

A similar approach was used when both knee angles and sEMG signals were used to train the network. Each signal (knee angle and sEMG) was split into 20-sample windows. A chronological sequence of 200-sample vectors was created, where each vector included the ten synchronized 20-sample windows from the sEMG signals of the eight muscles (four for each leg) and two knee-angle signals. In details, the first sample of the first 200-sample vector of the sequence was the first sample of the knee angle measured in the right leg; the second sample of the first 200-sample vector was the first sample of the EMG signal from the muscle 1 (TA, right leg), and so on up to the 10th signal (MH, left leg). Then a specific label was assigned to each 200-sample vector as follows: if the value of all the samples of the basographic signal corresponding to the 200-sample vector was 0 (or 1), a global label 0 (or 1) was assigned to the 200-sample vector. 200-sample vectors including swing-to-stance or stance-to-swing transitions were discarded. This approach including both Knee-angle and sEMG data to feed the neural network is referred to as KEMG approach.

3.4.3. Reference Approach

As reference experiment for direct comparison in the same population, a recent sEMG-based approach was used [14]. In this reference approach only sEMG signals were used to train the network. Similarly to the KEMG approach, each sEMG signal was split into 20-sample windows. A chronological sequence of 160-sample vectors was created, where each vector included the eight synchronized 20-sample windows from the sEMG signals of the eight muscles (four for each leg). In details, the first sample of the first 160-sample vector of the sequence was the first sample of the sEMG signal from the muscle 1 (TA, right leg), the second sample of the first 160-sample vector was the first sample of the EMG signal from the muscle 2 (GL, right leg), and so on up to the eighth signal (MH, left leg). Labeling was performed as in the previous approaches. This approach is referred to as the 'Reference approach'. As reported in [14], this approach has been previously validated versus a feature-based method, described in [13].

3.5. Training the Classifier

The present approach is based on the attempt at training the neural network classifier by means of sEMG data from 22 subjects out of 23 subjects of the present population (Learned set, LS) and then classifying gait phases in the remaining unseen subject (Unlearned set, US), following the so-called leave-one-out cross validation procedure. To this aim, all the vectors were picked up from the signals of the 22 subjects and then provided as input to the neural network for the training phase. The vectors from the remaining single subject were used for the testing phase, considering the corresponding foot-switch signal as ground truth. The procedure was performed twenty-three times, each time using a different subject as test set (23 folds cross-validation). For measuring the classification performances also for learned subjects, the set was split into training set (LS-train) and test set (LS-test). In details, LS-train includes the first 90% of each subject strand (approximately 3 min and 30 s, 180 gait cycles) and LS-test the remaining 10% (approximately 30 s, 20 gait cycles). Results in each subject were provided as the classification results in a single fold. Population (global) results were provided as mean value (± SD) over the 23 folds.

3.6. Neural Network

Multi-layer perceptron (MLP) architecture was implemented in the present study. The model was a deep neural network with three hidden layers composed of 512, 256, and 128 neurons and a one-dimensional output. The output was fed to a sigmoid function and a 0.5 threshold was used to achieve a binary output: when the output of the sigmoid was > 0.5 the label 1 was assigned, otherwise the label 0 was assigned. Rectified linear units (ReLU) were implemented to provide non-linearity between two consecutive hidden layers. In the experiments, stochastic gradient descent was employed as the optimization algorithm and binary cross entropy as the loss function. Eventually, MLP model was trained adopting an early stop technique: the network was trained for a maximum of 100 epochs, stopping when the accuracy on the validation set did not increase for 10 consecutive epochs. The best-performing learned parameters were adopted to evaluate the model performances.

3.7. Gait-Event Identification

The foot–floor-contact signal was predicted by chronologically arranging the binary output of MLP network. A vector was provided as output, where sequences of 1 (swing phase) alternate with sequences of 0 (stance phase). Literature reported that stance and swing phase during healthy walking at typical speed last on average around 60% and 40% of gait cycle. Starting from this observation, the predicted foot–floor-contact signal was cleaned by removing the sequences of samples shorter than 500 samples (~ 23% of gait cycle). Then, gait events were identified in the cleaned signal. Swing-to-stance transitions (heel strike, HS) were assessed as the sample when the sample value switched from 1 to 0. In the same way, stance-to-swing transitions (toe off, TO) were assessed as the sample when the sample value switched from 0 to 1. Performance of predictions was provided in terms of precision, recall, and F1-score.

A predicted HS or TO at time t_p was acknowledged as true positive (TP) if an event of the same type occurs in the ground truth signal at time t_g such that $|t_g - t_p| < T$. T is a temporal tolerance, set to 600 milliseconds. Otherwise, the predicted event was acknowledged as a false positive (FP). For all the true positives, mean absolute error (*MAE*) was computed as the average time distance between the predicted event and the corresponding one in ground truth signal.

3.8. Statistics

Shapiro–Wilk test was used to evaluate the hypothesis that each data vector had a normal distribution. Comparison between two normally distributed samples was performed with two-tailed, non-paired Student's *t*-test. The analysis of variance (ANOVA), followed by multiple comparison test, was used to compare more than two normally distributed samples. Kruskal–Wallis test was used to compare not normally distributed samples. Statistical significance was set at 5%.

4. Results

Average classification accuracies in every fold for the Knee, KEMG, and Reference approaches are shown in Table 1 for the learned-test set (LS-test) and in Table 2 for unlearned set (US). Figure 3 depicts a direct comparison of accuracy provided in each fold by Knee (red bars) vs. KEMG approach (blue bars). The direct comparison between mean values (horizontal dashed lines) shows a significant improvement of 4 points (94.6 ± 2.3% vs. 90.6 ± 2.9%, $p < 0.05$) of the classification accuracy provided by KEMG approach, compared with Knee approach. Starting from stance vs. swing classification, the present study is able to predict also the signal of foot–floor contact and to estimate gait events. An example of predictions of foot–floor-contact signal provided by Knee vs KEMG approaches is depicted in Figure 4. Tables 3 and 4 report the performance in US of HS and TO prediction in terms of mean absolute error (MAE), precision, recall, and F1-score. MAE detected in the prediction of HS provided by the Knee approach is significantly higher than MAE assessed by KEMG and Reference approaches (29.4 ± 13.7 ms vs. 18.8 ± 7.9 ms and 21.6 ± 7.0 ms; $p < 0.05$). Similarly, a significant higher MAE is observed in TO prediction provided by Knee approach (99.5 ± 28.9 ms vs. 35.9 ± 20.6 ms and 38.1 ± 14.2 ms; $p < 0.05$). No further significant differences were detected between groups.

Table 1. Stance vs. swing classification accuracy provided in LS-test (learned-test set).

Classification accuracy in LS-test (%)			
Fold	**Knee**	**KEMG**	**Reference**
1	90.7	95.6	94.9
2	90.8	95.1	94.6
3	90.8	95.1	94.4
4	91.3	96.0	94.8
5	91.2	95.6	95.0
6	90.9	95.4	94.9
7	90.8	95.7	95.0
8	90.8	95.5	94.7
9	90.5	95.8	94.8
10	90.4	95.2	94.6
11	91.3	96.0	95.0
12	89.9	95.9	94.8
13	91.4	95.9	94.8
14	91.5	95.2	95.0
15	90.6	95.8	94.8
16	91.2	95.5	94.7
17	91.0	95.6	94.7
18	90.2	95.5	94.8
19	90.8	95.4	94.8
20	90.7	95.5	94.8
21	91.8	96.3	95.3
22	91.5	95.8	95.0
23	90.8	95.6	94.8
Mean ± SD	90.9 ± 0.4	95.6 ± 0.3	94.8 ± 0.2

Table 2. Stance vs. swing classification accuracy provided in US

Classification Accuracy in US (%)			
Fold	**Knee**	**KEMG**	**Reference**
1	88.4	97.8	95.4
2	89.3	94.0	91.8
3	94.5	94.8	93.1
4	86.9	88.5	90.0
5	93.3	95.9	93.1
6	92.0	94.2	92.5
7	93.6	95.2	95.3
8	82.8	94.9	90.3
9	90.8	93.3	93.5
10	89.5	92.5	93.0
11	90.4	92.5	91.5
12	84.0	92.8	92.6
13	92.2	95.5	87.6
14	90.9	94.9	94.5
15	93.5	95.3	93.3
16	92.4	96.5	95.8
17	91.8	93.4	94.5
18	90.0	98.1	96.1
19	90.5	96.7	96.0
20	94.9	97.5	97.3
21	89.6	91.5	90.6
22	91.9	92.5	94.3
23	90.1	96.8	96.3
Mean ± SD	90.6 ± 2.9	94.6 ± 2.3	93.4 ± 2.4

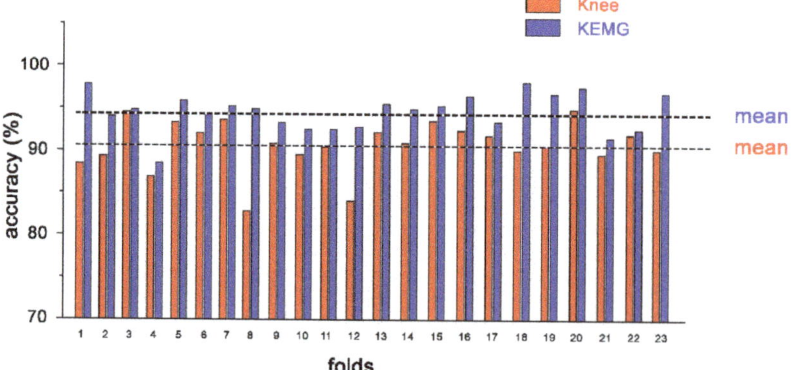

Figure 3. Direct comparison of classification accuracy provided in each fold by Knee (red bars) vs. KEMG approach (blue bars). Average values over 23 folds are represented with horizontal dashed lines.

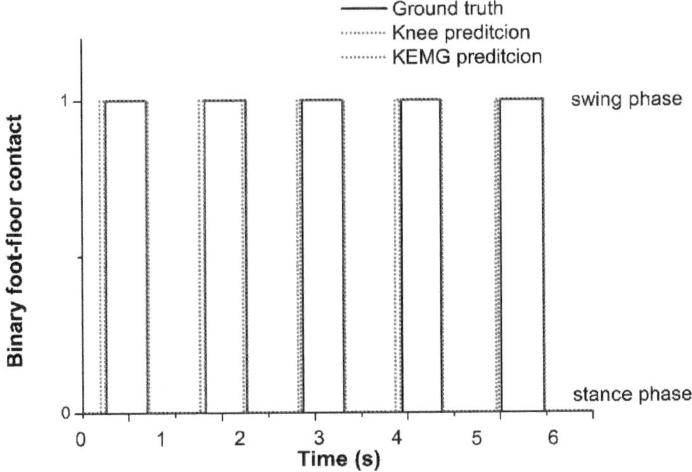

Figure 4. Example of predictions of foot–floor-contact signal in the same five strides of a representative subject, achieved by Knee (red dotted line) and KEMG (blue dotted line) approaches. Predictions are compared with the ground truth (black solid line).

Table 3. MAE (mean absolute error), precision, recall, and F1-score provided by Knee, KEMG, and reference (Ref) approach for heel strike (HS) prediction.

HS	MAE (ms)			Precision			Recall			F1-score		
Fold	**Knee**	**KEMG**	**Ref**	**Knee**	**KEMG**	**Ref**	**Knee**	**KEMG**	**Ref**	**Knee**	**KEMG**	**Ref**
1	31.8	9.9	22.3	1.00	1.00	0.99	1.00	1.00	0.99	1.00	1.00	0.99
2	38.4	34.0	31.3	1.00	1.00	0.99	0.99	0.98	0.97	1.00	0.99	0.98
3	13.4	20.6	29.4	1.00	0.99	0.99	1.00	0.99	0.99	1.00	0.99	0.99
4	27.9	15.0	16.0	1.00	1.00	1.00	0.95	1.00	1.00	0.97	1.00	1.00
5	17.4	14.5	25.0	1.00	1.00	1.00	0.99	1.00	1.00	0.99	1.00	1.00
6	18.2	16.4	22.5	1.00	1.00	1.00	0.97	1.00	1.00	0.99	1.00	1.00
7	15.2	25.2	25.0	1.00	1.00	1.00	1.00	1.00	1.00	1.00	1.00	1.00
8	71.1	24.7	25.2	1.00	1.00	0.99	0.99	1.00	0.99	1.00	1.00	0.99
9	53.0	20.2	19.4	1.00	1.00	1.00	0.98	1.00	0.99	0.99	1.00	0.99
10	28.5	30.4	38.2	1.00	1.00	1.00	0.99	0.99	0.99	0.99	0.99	0.99
11	17.6	9.6	14.3	1.00	1.00	1.00	0.98	0.98	0.97	0.99	0.99	0.99
12	25.2	7.1	10.6	1.00	1.00	1.00	1.00	1.00	1.00	1.00	1.00	1.00
13	31.6	18.2	29.0	1.00	1.00	1.00	0.94	0.98	0.86	0.97	0.99	0.92
14	47.9	15.5	18.0	1.00	1.00	1.00	0.73	0.99	0.98	0.85	1.00	0.99
15	33.1	17.2	21.8	1.00	1.00	1.00	0.98	1.00	1.00	0.99	1.00	1.00
16	29.6	15.9	18.7	1.00	1.00	1.00	1.00	1.00	1.00	1.00	1.00	1.00
17	38.8	37.0	27.0	1.00	1.00	1.00	1.00	1.00	1.00	1.00	1.00	1.00
18	29.7	9.6	13.1	1.00	1.00	1.00	1.00	1.00	1.00	1.00	1.00	1.00
19	26.5	19.6	12.7	1.00	1.00	1.00	1.00	1.00	1.00	1.00	1.00	1.00
20	14.0	10.4	13.4	1.00	1.00	1.00	1.00	1.00	1.00	1.00	1.00	1.00
21	18.5	20.3	22.4	0.94	0.99	0.98	0.96	0.97	0.98	0.95	0.98	0.98
22	21.2	25.7	26.6	1.00	1.00	1.00	1.00	1.00	0.99	1.00	1.00	1.00
23	28.8	15.0	14.8	0.99	1.00	1.00	0.99	1.00	1.00	0.99	1.00	1.00
Mean	29.4	18.8 *	21.6	1.00	1.00	1.00	[illegible]	[illegible]	0.99	0.99	1.00	0.99
SD	13.7	7.9	7.0	0.01	0.01	0.01	0.04	0.01	0.03	0.03	0.01	0.02

* means $p < 0.05$ between Knee and KEMG approach.

Table 4. MAE (mean absolute error), precision, recall, and F1-score provided by Knee, KEMG, and reference (Ref) approach for toe off (TO) prediction.

TO	MAE (ms)			Precision			Recall			F1-score		
Fold	Knee	KEMG	Ref	Knee	KEMG	Ref	Knee	KEMG	Ref	Knee	KEMG	Ref
1	116.4	16.8	26.6	1.00	1.00	1.00	1.00	1.00	1.00	1.00	1.00	1.00
2	126.8	23.4	40.1	0.99	0.99	0.96	0.98	0.97	0.94	0.98	0.98	0.95
3	125.3	31.9	22.6	1.00	0.98	0.99	1.00	0.99	1.00	1.00	0.98	0.99
4	99.6	104.0	80.8	0.99	0.99	1.00	0.93	0.99	0.99	0.96	0.99	0.99
5	99.6	28.6	30.1	1.00	1.00	1.00	0.98	1.00	0.99	0.99	1.00	1.00
6	56.4	47.1	47.4	1.00	1.00	1.00	0.97	1.00	1.00	0.99	1.00	1.00
7	97.9	33.2	31.2	1.00	1.00	1.00	1.00	1.00	1.00	1.00	1.00	1.00
8	170.5	18.7	54.1	0.96	0.99	0.97	0.95	0.99	0.97	0.96	0.99	0.97
9	67.9	50.6	44.2	1.00	1.00	1.00	0.97	0.99	0.98	0.98	1.00	0.99
10	87.0	51.6	34.0	1.00	1.00	0.99	1.00	1.00	0.99	1.00	1.00	0.99
11	107.8	53.3	59.0	0.93	0.94	0.96	0.91	0.93	0.93	0.92	0.93	0.94
12	67.6	64.7	59.1	1.00	1.00	1.00	0.99	1.00	1.00	0.99	1.00	1.00
13	76.7	20.6	39.9	1.00	1.00	0.98	0.93	0.98	0.84	0.96	0.99	0.90
14	72.3	33.5	26.4	1.00	1.00	1.00	0.73	1.00	0.98	0.85	1.00	0.99
15	59.4	37.7	55.0	0.99	1.00	1.00	0.97	1.00	1.00	0.98	1.00	1.00
16	135.8	20.2	30.0	1.00	1.00	1.00	1.00	1.00	1.00	1.00	1.00	1.00
17	94.5	32.6	28.2	1.00	1.00	1.00	1.00	1.00	1.00	1.00	1.00	1.00
18	119.4	12.3	23.9	1.00	1.00	1.00	1.00	1.00	1.00	1.00	1.00	1.00
19	122.1	17.8	30.5	0.99	1.00	1.00	0.99	1.00	0.99	0.99	1.00	0.99
20	119.2	19.1	17.6	1.00	1.00	1.00	1.00	1.00	1.00	1.00	1.00	1.00
21	90.9	38.3	37.6	0.96	0.98	0.96	0.97	0.95	0.95	0.96	0.96	0.96
22	117.4	49.5	35.2	0.97	0.98	0.99	0.97	0.98	0.98	0.97	0.98	0.99
23	59.5	21.3	23.8	1.00	1.00	1.00	1.00	1.00	1.00	1.00	1.00	1.00
Mean	99.5	35.9*	38.1	0.99	0.99	0.99	0.97	0.99	0.98	0.98	0.99	0.98
SD	28.9	20.6	14.2	0.01	0.01	0.01	0.05	0.01	0.04	0.04	0.01	0.02

* means $p < 0.05$ between Knee and KEMG approach.

5. Discussion

The goal of this study is to propose a novel approach for classifying stance vs. swing and assessing HS and TO timing, based on deep learning analysis of sagittal knee-angle data measured with a single electrogoniometer per each leg. This so-called Knee approach achieves average stance/swing-classification accuracy over 23 folds (± SD) of 90.9 ± 0.4% in LS-test and 90.6 ± 2.9% in US (last row of Tables 1 and 2, respectively). A reduction of accuracy is detected, compared to Reference approach in both LS-test (3.9 points) and US (2.8 points). This reduction is expected, since only one signal per leg is used in Knee approach vs. the four signals per leg used in the Reference approach (more input information, better classification performance). Despite this, the average accuracy of stance/swing classification is still > 90% and falls in the range identified by the different machine-learning-based approaches (sEMG, angular sensors) reported in literature [10,13,21,23,24] (see Section 2). Moreover, the absence of any significant difference between classification accuracies in US vs. LS-test ($p > 0.05$) highlights that the network is able to keep the same performance even when tested on brand new subjects (US). Classification accuracy > 90% in US subjects is supposed to be very useful in clinical environments, where brand new subjects are analyzed every day. This outcome is associated also to a limited standard deviation, as in the Reference approach. As expected, SD is higher in US, indicating a large variability of classification for subjects not used during training phase.

Besides the suitable classification performance, a reliable post-processing of model output was implemented for gait-event estimation in US (see Section 3.7), ensuring values of prediction, recall, and F1-score very close to 1 (Tables 3 and 4). These values are not statistically different from the correspondent values provided by the Reference approach. Furthermore, a mean MAE over population of 29.4 ± 13.7 ms and 99.5 ± 28.9 ms is achieved in predicting HS and TO (Tables 3 and 4, respectively).

Compared to Reference value, average MAE value in HS prediction is 7.8 ms higher. However, Knee approach performs better than the sEMG approach proposed in [13], which achieved a mean HS MAE of 35 ms. TO prediction is less accurate: mean *MAE* value of 99.5 ms vs. 38.1 ms (Reference value, Table 4). It has been reported that it is more challenging identifying toe-offs rather than heel-strikes [13,14,33]. Thus, higher MAE in TO prediction was expected. Liu et al. achieved high accuracy in classifying gait phases with joint-angular-sensor data [10]. However, despite not reporting detailed MAE for TO prediction, they detected the most relevant recognition errors just around the transition from stance to swing (i.e., TO). Differences in toe-off MAE compared to above-mentioned studies would be likely attributable to a different number of signals (and consequently of sensors) used in the different approaches: one single signal per leg in Knee method, two signals in [13], four signals in the Reference approach, and even more in [10]. Thus, the desirable simplification of experimental set-up (one single sensor) is paid with a deterioration of only TO (not HS) prediction. However, this could be a good compromise for general task such as stride recognition, stride-time computation, identification of toe walking, and so on, where only HS event is involved. Moreover, it should be taken into account that present performances are achieved in condition of high variability of foot–floor contact, due to the eight-shaped path followed during ground (not treadmill) walking which includes acceleration, deceleration, curves, and reversing. Larger variability of the signal to predict, indeed, is expected to affect the performance of the classifier.

As mentioned above, promising performances in classifying gait phases and predicting gait events are provided by studies proposing a machine learning analysis of only sEMG signals, [13,14,21,23,24]. All those studies present suitable and reliable outcomes, but, to our knowledge, the best results in terms of mean absolute value in the prediction of HS and TO are achieved in a recent study of the present group of studies [14]. The technique introduced by this study is adopted here as the Reference approach. The present study is further aimed to test if the addition of electrogoniometer data could improve the classification performance of this Reference approach. The approach including both knee-angle and sEMG data to feed the neural network is referred to as KEMG approach. Detailed accuracy values in the 23 folds for stance/swing classification accomplished in LS-test and US are shown in Table 1 and in Table 2, respectively. Comparison analysis (in Figure 3 for US) shows as KEMG approach (blue bars) achieves improved accuracy values in each one of the 23 folds, with respect to Knee approach (red bars), implying a significant increase (around 4 points for both LS-test and US, $p < 0.05$) of mean accuracy over 23 folds. Mean accuracies of KEMG approach outperform of around 1 point also the Reference approach: i.e., 95.6 ± 0.3% vs. 94.8 ± 0.2% in Learned set and 94.6 ± 2.3% vs. 93.4 ± 2.3% in Unlearned set. As for Knee and Reference approaches, KEMG provides values of prediction, recall, and F1-score very close to 1, in predicting HS and TO (Tables 3 and 4). MAE values are significantly lower than the correspondent values provided by Knee approach (reduction of 36.1% for HS and 63.9% for TO, $p < 0.05$). It is worth noticing that also SD values decreased (from 13.7 to 7.9 for HS and from 28.9 to 20.6 for TO), suggesting an improved repeatability of prediction quality among different folds. A significant improvement of prediction error in KEMG is observed also compared to Reference approach, in terms of reduction of MAE (18.8 vs. 21.6 for HS and 35.9 vs. 38.1 for TO). These outcomes suggest that the introduction of knee-angle data could improve the performances of sEMG-based approaches, both in classification accuracy and in prediction error.

As introduced earlier, the clinically oriented aim of this work is trying to simplify the experimental set-up associated to instrumental gait analysis, assessing the signal of foot–floor contact from deep learning analysis of sagittal knee-angles measured by a single electrogoniometer. Gait analysis is acknowledged as a suitable procedure for quantitatively estimating the deterioration of motor function in clinics. The issue of cumbersome and time-consuming experimental protocols is getting increasingly relevant, particularly for evaluation in pathology. This is true for classical approaches based on foot-switch sensors, pressure mats, and stereo-photogrammetric systems, but also for the more recently-developed wearable sensors, which could need specific care for the suitable placement and necessity of precise calibration process, not always compatible with the clinical timetable.

Thus, an approach based on a single, reliable, easy-to-attach sensor (electrogoniometer) is truly desirable: the fewer sensors are involved, the simpler is to protect patient comfort. Aforesaid studies seem to indicate that a large data-set of signals from many sensors is needed to classify gait phases and/or estimate gait events during normal or aided walking [10,13,14]. Outcomes achieved here suggest that it is strongly dependent on the task to pursue. If the aim is to classify stance vs. swing phase or to assess gait parameters where only HS event is involved (stride recognition, stride-time computation, identification of toe walking), the present Knee approach provides performances in line with what reported in literature, but with the clinical advantage of using one single simple sensor. When the aim is more complicated to pursue (gait sub-phase recognition, swing and stance time duration and so on) and/or more elevated performances are needed, approaches based on sensor fusion, as KEMG approach proposed here, are preferable. Thus, the main contribution of the present study consists in showing that for specific simple (but essential) tasks such as stride recognition, stride-time computation, and identification of toe walking, the single-sensor approach is able to provide classification performance comparable to those achieved by multi-sensor approaches. The information included in the present study would be particularly suitable for specific environments, such as the walking-aid devices or of portable rehabilitation system [34–37], where sensors could already be embedded in the system.

6. Conclusions

The present study proposes a novel methodology for classifying stance vs. swing and predicting gait-event timing, based on neural-network classification of signals acquired by a single knee electrogoniometer during walking. The clinically useful contribution of the study consists in assessing gait events from only sagittal knee-angle signals, avoiding the installation of additional sensors on the human body and promoting the reduction of the sensor-system complexity. Additional goal is to evaluate if the introduction of knee-angle data from the electrogoniometer could improve the classification performance of state-of-the-art sEMG-based methods, in order to provide a sensor-fusion approach useful to face more complex task or to pursue higher classification/prediction performances. The comparison of the two approaches shows as the reduction of set-up complexity implies a worsening of classification performances. However, the choice of the suitable approach should not only be driven by network performance but also (mainly) by patient comfort and clinical needs.

Author Contributions: Conceptualization, F.D.N., C.M., S.F., and A.C.; Methodology, F.D.N. and C.M.; Software, C.M.; Investigation, C.M. and F.D.N.; Validation, C.M.; Resources, S.F.; Data curation, F.D.N. and C.M.; Writing—original draft preparation, F.D.N.; Writing—review and editing, A.C. and S.F.; Visualization, F.D.N. and C.M.; Supervision, A.C. and S.F. All authors have read and agreed to the published version of the manuscript.

Funding: This research received no external funding.

Conflicts of Interest: The authors declare no conflict of interest.

References

1. Pappas, I.P.I.; Popovic, M.R.; Keller, T.; Dietz, V.; Morari, M. A reliable gait phase detection system. *IEEE Trans. Neural Syst. Rehabil. Eng.* **2001**, *9*, 113–125. [CrossRef] [PubMed]
2. Rueterbories, J.; Spaich, E.G.; Larsen, B.; Andersen, O.K. Methods for gait event detection and analysis in ambulatory systems. *Med. Eng. Phys.* **2010**, *32*, 545–552. [CrossRef] [PubMed]
3. Qi, Y.; Soh, C.B.; Gunawan, E.; Low, K.; Thomas, R. Assessment of foot trajectory for human gait phase detection using wireless ultrasonic sensor network. *IEEE Trans. Neural Syst. Rehabil.* **2016**, *24*, 88–97. [CrossRef] [PubMed]
4. Zheng, E.; Chen, B.; Wei, K.; Wang, Q. Lower limb wearable capacitive sensing and its applications to recognizing human gaits. *Sensors (Basel)* **2013**, *13*, 13334–13355. [CrossRef] [PubMed]
5. Grimmer, M.; Schmidt, K.; Duarte, J.E.; Neuner, L.; Koginov, G.; Riener, R. Stance and Swing Detection Based on the Angular Velocity of Lower Limb Segments during Walking. *Front. Neurorobot.* **2019**, *24*, 13–57. [CrossRef] [PubMed]

6. Miller, A. Gait event detection using a multilayer neural network. *Gait Posture* **2009**, *29*, 542–545. [CrossRef]
7. Taborri, J.; Palermo, E.; Rossi, S.; Cappa, P. Gait Partitioning Methods: A Systematic Review. *Sensors (Basel)* **2016**, *16*, 66. [CrossRef]
8. Novak, D.; Gorsic, M.; Podobnik, J.; Munih, M. Toward real-time automated detection of turns during gait using wearable inertial measurement units. *Sensors (Basel)* **2014**, *14*, 18800–18822. [CrossRef]
9. Osis, S.T.; Hettinga, B.A.; Ferber, R. Predicting ground contact events for a continuum of gait types: An application of targeted machine learning using principal component analysis. *Gait Posture* **2016**, *46*, 86–90. [CrossRef]
10. Liu, D.X.; Wu, X.; Du, W.; Wang, C.; Xu, T. Gait Phase Recognition for Lower-Limb Exoskeleton with Only Joint Angular Sensors. *Sensors (Basel)* **2016**, *16*, 1579. [CrossRef]
11. Pacini Panebianco, G.; Bisi, M.C.; Stagni, R.; Fantozzi, S. Analysis of the performance of 17 algorithms from a systematic review: Influence of sensor position, analysed variable and computational approach in gait timing estimation from IMU measurements. *Gait Posture* **2018**, *66*, 76–82. [CrossRef] [PubMed]
12. Kidziński, Ł.; Delp, S.; Schwartz, M. Automatic real-time gait event detection in children using deep neural networks. *PLoS ONE* **2019**, *14*, e0211466. [CrossRef] [PubMed]
13. Nazmi, N.; Abdul Rahman, M.; Yamamoto, S.I.; Ahmad, S. Walking gait event detection based on electromyography signals using artificial neural network. *Biomed. Signal Process. Control* **2019**, *47*, 334–343. [CrossRef]
14. Morbidoni, C.; Cucchiarelli, A.; Fioretti, S.; Di Nardo, F. A Deep Learning Approach to EMG-Based Classification of Gait Phases during Level Ground Walking. *Electronics* **2019**, *8*, 894. [CrossRef]
15. Mohammed, S.; Same, A.; Oukhellou, L.; Kong, K.; Huo, W.; Amirat, Y. Recognition of gait cycle phases using wearable sensors. *Robot. Auton. Syst.* **2016**, *75*, 50–59. [CrossRef]
16. Morettini, M.; Peroni, C.; Sbrollini, A.; Marcantoni, I.; Burattini, L. Classification of drug-induced hERG potassium-channel block from electrocardiographic T-wave features using artificial neural networks. *Ann. Noninvasive Electrocardiol.* **2019**, *24*, e12679. [CrossRef]
17. Di Nardo, F.; Mengarelli, A.; Ghetti, G.; Fioretti, S. Statistical analysis of EMG signal acquired from tibialis anterior during gait. *IFMBE Proc.* **2014**, *41*, 619–622. [CrossRef]
18. Gurney, J.; Kersting, U.; Rosenbaum, D. Between-day reliability of repeated plantar pressure distribution measurements in a normal population. *Gait Posture* **2008**, *27*, 706–709. [CrossRef]
19. Bovi, G.; Rabuffetti, M.; Mazzoleni, P.; Ferrarin, M. A multiple-task gait analysis approach: Kinematic, kinetic and EMG reference data for healthy young and adult subjects. *Gait Posture* **2011**, *33*, 6–13. [CrossRef]
20. Tang, Z.; Zhang, K.; Sun, S.; Gao, Z.; Zhang, L.; Yang, Z. An upper-limb power-assist exoskeleton using proportional myoelectric control. *Sensors (Basel)* **2014**, *14*, 6677–6694. [CrossRef]
21. Joshi, C.D.; Lahiri, U.; Thakor, N.V. Classification of gait phases from lower limb EMG: Application to exoskeleton orthosis. In Proceedings of the 2013 IEEE Point-of-Care Healthcare Technologies (PHT), Bangalore, India, 16–18 January 2013; pp. 228–231. [CrossRef]
22. Jung, J.; Heo, W.; Yang, H.; Park, H. A neural network-based gait phase classification method using sensors equipped on lower limb exoskeleton robots. *Sensors (Basel)* **2015**, *15*, 27738–27759. [CrossRef] [PubMed]
23. Ziegier, J.; Gattringer, H.; Mueller, A. Classification of Gait Phases Based on Bilateral EMG Data Using Support Vector Machines. In Proceedings of the IEEE RAS and EMBS International Conference on Biomedical Robotics and Biomechatronics, Enschede, The Netherlands, 26–29 August 2018; pp. 978–983. [CrossRef]
24. Meng, M.; She, Q.; Gao, Y.; Luo, Z. EMG signals based gait phases recognition using hidden Markov models. In Proceedings of the 2010 IEEE International Conference on Information and Automation, ICIA 2010, Harbin, China, 20–23 June 2010; pp. 852–856. [CrossRef]
25. Morbidoni, C.; Principi, L.; Mascia, G.; Strazza, A.; Verdini, F.; Cucchiarelli, A.; Di Nardo, F. Gait Phase Classification from Surface EMG Signals Using Neural Networks. *IFMBE Proc.* **2020**, *76*, 75–82. [CrossRef]
26. Agostini, V.; Balestra, G.; Knaflitz, M. Segmentation and Classification of Gait Cycles. *IEEE Trans. Neural Syst. Rehabil. Eng.* **2014**, *22*, 946–952. [CrossRef] [PubMed]
27. Catalfamo, P.; Moser, D.; Ghoussayni, S.; Ewins, D. Detection of gait events using an F-Scan in-shoe pressure measurement system. *Gait Posture* **2008**, *28*, 420–426. [CrossRef] [PubMed]
28. Crea, S.; de Rossi, S.M.M.; Donati, M.; Reberšek, P.; Novak, D.; Vitiello, N.; Lenzi, T.; Podobnik, J.; Munih, M.; Carrozza, M.C. Development of gait segmentation methods for wearable foot pressure sensors. In Proceedings

of the 34th IEEE Engineering in Medicine and Biology Society (EMBS), San Diego, CA, USA, 28 August–1 September 2012; pp. 5018–5021. [CrossRef]

29. Senanayake, C.M.; Senanayake, S.M.N.A. Computational intelligent gait-phase detection system to identify pathological gait. *IEEE Trans. Inf. Technol. Biomed.* **2010**, *14*, 1173–1179. [CrossRef] [PubMed]
30. González, I.; Fontecha, J.; Hervás, R.; Bravo, J. An ambulatory system for gait monitoring based on wireless sensorized insoles. *Sensors (Basel)* **2015**, *15*, 16589–16613. [CrossRef]
31. Hermens, H.J.; Freriks, B.; Merletti, R.; Stegeman, D.; Blok, J.; Rau, G.; Disselhorst-Klug, C.; Hägg, G. European recommendations for surface electromyography, SENIAM. *Roessingh Res. Dev.* **1999**, *8*, 13–54.
32. Winter, D.A.; Yack, H.J. EMG profiles during normal human walking: Stride-to-stride and inter-subject variability. *Electroencephalogr. Clin. Neurophysiol.* **1987**, *67*, 402–411. [CrossRef]
33. Khandelwal, S.; Wickstrasm, N. Evaluation of the performance of accelerometer-based gait event detection algorithms in different real-world scenarios using the MAREA gait database. *Gait Posture* **2017**, *51*, 84–90. [CrossRef] [PubMed]
34. Kazerooni, H.; Steger, R.; Huang, L.H. Hybrid control of the Berkeley Lower Extremity Exoskeleton (BLEEX). *Int. J. Robot. Res.* **2006**, *25*, 561–573. [CrossRef]
35. Ma, Y.; Xie, S.; Zhang, Y. A patient-specific EMG-driven neuromuscularmodel for the potential use of human-inspired gait rehabilitation robots. *Comput. Biol. Med.* **2016**, *70*, 88–98. [CrossRef] [PubMed]
36. Azimi, V.; Nguyen, T.T.; Sharifi, M.; Fakoorian, S.A.; Simon, D. Robust Ground Reaction Force Estimation and Control of Lower-Limb Prostheses: Theory and Simulation. *IEEE Trans. Syst. Man Cybern. Syst.* **2018**, *99*, 1–12. [CrossRef]
37. Watanabe, T.; Endo, S.; Morita, R. Development of a prototype of portable FES rehabilitation system for relearning of gait for hemiplegic subjects. *Healthc. Technol. Lett.* **2016**, *3*, 284–289. [CrossRef] [PubMed]

Article

Estimation of Knee Movement from Surface EMG Using Random Forest with Principal Component Analysis

Zhong Li, Xiaorong Guan *, Kaifan Zou and Cheng Xu

School of Mechanical Engineering, Nanjing University of Science and Technology, Nanjing 210094, China; zhong0814@njust.edu.cn (Z.L.); zoukaifan@njust.edu.cn (K.Z.); xucheng62@njust.edu.cn (C.X.)

* Correspondence: gxr@njust.edu.cn; Tel.: +86-186-5209-6966

Received: 28 November 2019; Accepted: 25 December 2019; Published: 28 December 2019

Abstract: To study the relationship between surface electromyography (sEMG) and joint movement, and to provide reliable reference information for the exoskeleton control, the sEMG and the corresponding movement of the knee during the normal walking of adults have been measured. After processing the experimental data, the estimation model for knee movement from sEMG was established using the novel method of random forest with principal component analysis (RFPCA). The influence of the sample size and the previous sEMG data on the prediction efficiency was analyzed. The estimation model was not sensitive to the sample size when samples increased to a certain value, and the results of different previous sEMG showed that the prediction accuracy of the estimation models did not always improve with the increasing features of input. By comparing the estimation model of back propagation neural network with principal component analysis (BPPCA), it was found that RFPCA was suitable for all participants in the experiment with less execution time, and the root mean square error was around 5° which was lower than BPPCA with errors varying from 7° to 25°. Therefore, it was concluded that the RFPCA method for the estimation of knee movement from sEMG is feasible and could be used for motion analysis and the control of exoskeleton.

Keywords: sEMG; knee; random forest; principal component analysis; back propagation; estimation model

1. Introduction

The exoskeleton robot could be a revolutionary technology in human limb rehabilitation [1] and power enhancement [2]. However, the motion intention of its wearer has limited the development of this technology, because traditional sensors for the exoskeleton are unable to detect motion tendency ahead of time. Since the surface electromyography (sEMG) signal is noninvasive, and has the potential to predict people's movement intentions 30–100 ms in advance [1], the sEMG has been favored by many researchers, and with the help of sEMG sensors, the performance of wearable devices would be improved [1,3–5]. Thus, in addition to the exoskeleton technology, there are a wide range of applications for using sEMG, such as wearable devices [6], prosthetic limbs [7], and other such myoelectric control systems [8]. The existing studies focus on how the sEMG signals relate to human movement for better control of the wearable devices and similar products.

Some of the existing research has investigated the relationship between sEMG and human biomechanics. Chen et al. [9] proposed a musculoskeletal biomechanical model connecting sEMG and knee joint torque, based on the underlying physiological mechanism facilitating the study of neural control. Tagliapietra et al. [10] used a subject-specific EMG-driven Neuro MusculoSkeletal (NMS) model to estimate ankle torque and muscle forces expressed by the subject. Zhuang et al. [11] proposed an sEMG-based admittance controller that could enable a more synchronized human–robot interaction, as compared to the torque-sensing-based admittance controller.

However, to avoid building a complicated biomechanical model, some researchers have tried to use a data training method to do the job. Anwar et al. [12] proposed an adaptive neuro fuzzy inference system (ANFIS), such as a neuro-fuzzy type knowledge-based adaptive network that contained a non-parametric model, with an EMG signal of two muscles used as the input to estimate torque. Gui et al. [13] used radial basis function (RBF) neural networks to approximate the active joint torque of subjects during the swing phase.

Besides the biomechanical method, there is a new intuitive myoelectric control strategy for assistive devices, which relies on the sEMG-based intention estimation of human motion. These predictions can be broadly categorized as classification and regression models [14]. For the classification, Toledo-Pérez et al. [15] used a support vector machine (SVM) based on sEMG to classify the intention of right lower limb movement. Morbidoni et al. [16] proposed a deep learning (DL) approach for sEMG-based classification of stance/swing phases and the prediction of the foot–floor-contact signal in more natural walking conditions. Nazmi et al. [17] proposed a classification system for both stance and swing phases, by extracting the patterns of electromyography signals from time domain features and feeding them into an artificial neural network (ANN) classifier.

For the continuous estimation of the joint angle, there are various methods. Bao et al. [18] presented a single stream convolutional neural network (CNN) for mapping sEMG to wrist angles within three degrees-of-freedoms. Xiao et al. [19] used the mean absolute value, waveform length, zero crossing, slope signs changes, and the difference in absolute standard deviation value of sEMG, in order to estimate continuous elbow motion by random forest (RF). Lei Z. [20] used the back propagation (BP) neural network to establish a model of the relationship between elbow angles and sEMG signals features, through which they estimated the angles of the elbow joint and achieved continuous motion control of the exoskeleton. Huang et al. [21] presented deep-recurrent neural networks (RNNs) for predicting the knee joint angle in real-time, based on a fusion of sEMG and kinematics signals.

It can be concluded that most of the existing studies are model-free approaches for the estimation of joint angles from sEMG, based on machine learning (ML). Furthermore, the majority of the existing research used a single method of ML, and most of them seldom considered the influence of the previous sEMG as the input on the accuracy of their methods, even though both the joint movement and sEMG signals are in a continuous time-sequence. Also, the size of the training sample for ML is also a debatable issue in terms of the accuracy of estimation, since a large training sample would lead to overfitting and be time consuming.

Thus, a novel double-ML-method, based on random forest combined with principal component analysis (RFPCA) has been proposed in this work to estimate the movement of the knee joint from sEMG, with the expectation of achieving high accuracy and efficiency. This method was also utilized to analyze how the input of the previous sEMG and the sample size for model validation affect the estimation of the knee joint movement. Moreover, a BPPCA was constructed to compare with the RFPCA, and the RFPCA presented a better performance in this work.

The remainder of the paper is organized as follows. Section 2 introduces the experiment and proposes the knee angle estimation method. Section 3 presents the estimation results using RFPCA and BPPCA, followed by the discussion in Section 4. Finally, the conclusions are drawn in Section 5.

2. Materials and Methods

2.1. Subjects

A total of six healthy subjects, who have never suffered from muscular atrophy or disorders, participated in this study, with an average height, age and weight of 181 ± 3.8 cm, 72.5 ± 6.9 kg and 24.2 ± 1.6 yrs, respectively. All of the subjects enrolled in this study knew the procedure of the experiment, and signed and agreed to participate in the experimental study as the test subjects.

The experiment was approved by Nanjing University of Science and Technology (Date granted: June 12, 2019) and performed in accordance with the Declaration of Helsinki.

2.2. Experimental Preparation and Protocol

The unilateral lower limb of the human body contains at least 30 muscles that drive 7 degrees of freedom of the lower limb. Since the knee is a vital joint in the lower limb, we have chosen it as the key aspect in our research. The sEMG signals related to the knee of the unilateral leg were obtained by Trigno wireless sEMG instrument. To obtain better sEMG signals, several muscles that are easily detected were selected, including the vastus lateralis (VL), rectus femoris (RF), vastus medialis (VM), gastrocnemius medialis (GM) and gastrocnemius lateralis (GL). The adhesive positions (shown in Figure 1) of the surface electrodes are the positions recommended by SENIAM [22]. Before the experiment, the skin surfaces, where the electrodes had to be placed, were shaved and then cleaned with alcohol. This was done to reduce the impedance between the measured skin and electrodes and also to improve the sensor–skin contact.

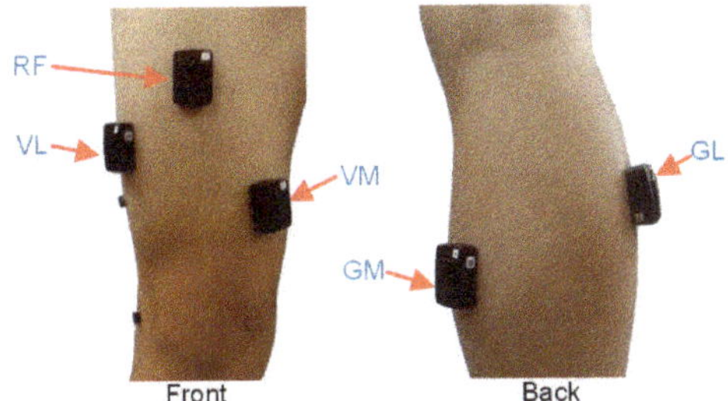

Figure 1. The locations of surface electromyography (sEMG) sensors.

On the lateral side of the leg with sEMG sensors, three markers were placed to obtained the kinematic data of the knee, through a 3D motion capture system called Codamotion. As shown in Figure 2, Marker B was placed at the approximate center of the knee joint, on the sagittal skin of the subject, while Marker A and Marker C were respectively placed on the projection line of the thigh femur and calf tibia on the sagittal skin of the subject. The approximate flexion–extension motion angle of the knee joint was obtained by collecting the spatial motion trajectory of Markers A, B and C.

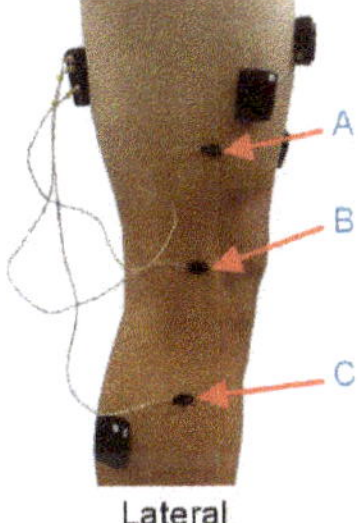

Figure 2. The placement of the Codamotion markers.

During the experiment, the subject was asked to walk back and forth nearly 20 times along a straight line about 5 m in length, at his natural or free cadence, as shown in Figure 3. The two Codamotion CX1 units on the right of the subject collected the kinematic data of the markers at a sampling frequency of 100 Hz, while the wireless Trigno system picked up the raw sEMG signals from the five muscles at a frequency of 2000 Hz. The data were then synchronized and transferred to the Codamotion hub, and finally, transmitted to the PC and stored. If the subject felt uncomfortable while the experiment was in progress, he would rest for 10 minutes to reduce the effects of such factors as fatigue. If he felt better after the rest, the experiment would continue. Otherwise, the experiment would be stopped. In this way, the data of one trail from as many as 8 strides could be collected, which was more than adequate to obtain 4 representative sEMG profiles for each muscle and gait cycle (GC) of the leg. Approximately 80 full GCs and their corresponding sEMG signals were recorded in all for each subject. Since each subject needed to be properly equipped with the sensors, and it required approximately one hour for each subject to undergo the experiment process, including preparation (15 min), resting (30 min) and walking (15 min), the entire experiment lasted 2 days, with one person in the morning and two in the afternoon.

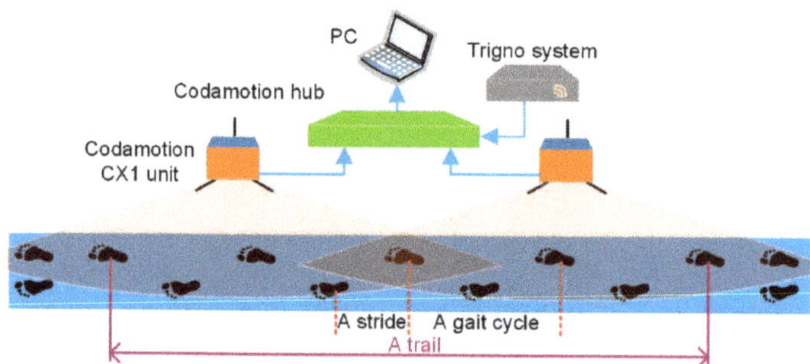

Figure 3. The schematic of the experiment.

2.3. Signal Processing

The sEMG signals were focused on the time domain because of the time-sequence. The amplitude extracted from the raw sEMG signals was used for the training of the RFPCA, similar to the work done by the authors of [16], who directly used the envelopes of the EMG signal to train the network. This feature was carried out from digital filtering and using simple math. A Butterworth band-pass filter (band length of 10–500 Hz, 4th order) was used to filter interference signals and extract effective signals. After that, a full-wave rectification of the sEMG signal was conducted, and the signal was then filtered through a low-pass filter (Butterworth at 6 Hz, 2nd order) [23]. One of the representative examples of the sEMG signal processing of one trial is shown in Figure 4.

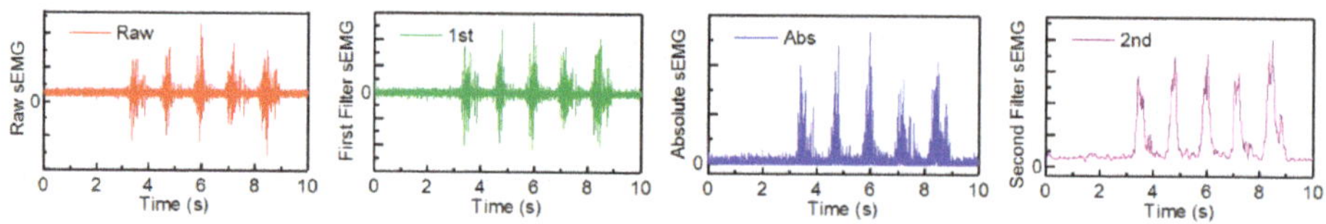

Figure 4. The signal processing of sEMG.

For processing the kinematic data of the knee, the vector $\vec{V}_{AB}$ consisting of Markers A and B represents the direction of the thigh, while vector $\vec{V}_{BC}$ consisting of Markers B and C represents the direction of the shank. The motion angle of knee could then be calculated approximatively as:

$$\theta_{knee} = \arccos\left(\frac{\vec{V}_{AB} \cdot \vec{V}_{BC}}{\left|\vec{V}_{AB}\right|\left|\vec{V}_{BC}\right|}\right). \tag{1}$$

In order to keep the data synchronized with the knee and facilitate subsequent knee movement prediction, the sEMG signal, after the original processing, was resampled at 100 Hz. Thus, a set of data $\left[E_{\mathrm{VL},t}, E_{\mathrm{RF},t}, E_{\mathrm{VM},t}, E_{\mathrm{GM},t}, E_{\mathrm{GL},t}, y_{knee,t}\right]$ at time t was obtained, where $E_{i,t}$ represents the envelope of i sEMG signal after processing, and $y_{knee,t}$ represents the knee angle θ_{knee} at time t.

For the study of the sEMG historical influence, the data from one trail of each subject was chosen for training, in order to establish the estimation models. The influence of previous sEMG signals was studied by the number of previous sEMG signals used as a feature vector in the learning course. A typical sample is a combination of sEMG and knee angle $[\boldsymbol{X}_t, Y_t]$, where the sEMG $\boldsymbol{X}_t$ is the input, while the knee angle $Y_t = y_{knee,t}$ is the output. In this work, the sEMG input $\boldsymbol{X}_t$ at time t is defined as follows:

$$\boldsymbol{X}_t = \left[\boldsymbol{x}_{t-n\Delta t}, \boldsymbol{x}_{t-(n-1)\Delta t}, \cdots, \boldsymbol{x}_{t-\Delta t}, \boldsymbol{x}_t\right], \tag{2}$$

where, $\boldsymbol{x}_t$ is the vector of sEMG at time t after processing, namely $[E_{\mathrm{VL},t}, E_{\mathrm{RF},t}, E_{\mathrm{VM},t}, E_{\mathrm{GM},t}, E_{\mathrm{GL},t}]$; n is the number of previous sEMG signals used as the input, Δt is the sample interval which is 1/100 Hz. The input dimension of $\boldsymbol{X}_t$ increases as n increases.

For the sample size, n was zero, and all of the processed data of each subject was segmented into groups according to the GC. One group contains different sampling numbers for different subjects, which are listed in Table 1. For each subject, 61 complete groups of data (GDs, approximately 7808 samples) were selected for the following work.

Table 1. The average sampling numbers contained in a gait cycle (GC) of different subjects.

Subject	**1#**	**2#**	**3#**	**4#**	**5#**	**6#**	**Mean**
Number	129 ± 9	127 ± 6	126 ± 7	131 ± 6	134 ± 9	122 ± 7	128 ± 8

2.4. Estimation Model

Because of the complexity of the sEMG signal and the differences between subjects, it is hard to establish a general mathematical model to represent the mapping relationship from sEMG to the knee angle. Furthermore, the biomechanical model describing the relationship between the sEMG and joint angle is also complicated and difficult to construct for practical application. Therefore, to establish a universal sEMG–angle model of a human joint, with a learning function, this study adopted a novel model-free method using random forest (RF) in combination with principal component analysis (PCA), in order to set up the estimation model between sEMG signal and knee movement. It is expected that this coupled ML method will be able to handle the estimation issue for different participants with a parametric adaptive approach. The input of the model is the processed sEMG, and the output is the knee angle.

2.4.1. Random Forest

Random forest (RF), based on the theory of decision trees, was first proposed by Breiman [24]. RF is an effective tool in prediction, because with the right input, RF produces accurate classifiers and regressors [24]. In standard trees, each node is split using the best split among all variables. In a RF, each node is split using the best among a subset of predictors randomly chosen at that node. This somewhat counterintuitive strategy performs very well in comparison to other classifiers, including discriminant analysis, SVM and neural networks, and is robust against overfitting [25]. Additionally, the execution time of RF is far less than RBF and SVM when used to process high dimensional data, because the RF algorithm itself can select the important features automatically [19]. It is also relatively robust to outliers and noise. Due to these advantages, RF was chosen for application to the relationship between the sEMG and knee joint in this study.

As an ensemble learning method, RF achieves better generalization performance by establishing multiple decision trees. If RF has N decision trees, it is necessary to generate N sample sets to train each tree. Each tree is grown as follows [26]:

- If the number of cases in the training set is T_r, then T_r cases at random are sampled—but with replacement, from the original data. This sample will be the training set for growing the tree.
- If there are I_v input variables, a number $N_f << I_v$ is specified, such that at each node, N_f variables are selected at random out of the I_v and the best split on these, m, is used to split the node. The value of N_f is held constant during the forest growing.
- Each tree is grown to the largest extent possible. There is no pruning.

2.4.2. Principal Component Analysis

Principal component analysis (PCA) is a statistical technique that performs a linear transformation from an original set of values into a smaller one of uncorrelated variables [15]. The idea was conceived of by K. Pearson [27] and later developed by Hotelling [28].

PCA needs to find some encoding function that produces the code for an input, and a decoding function that produces the reconstructed input, given its code [29]. In general, PCA uses a covariance matrix to reduce the data dimension. By calculating the eigenvalue eigenvector of the covariance matrix of the data, and selecting the matrix composed of eigenvectors corresponding to k features with the largest eigenvalue (i.e., the largest variance), the data matrix can be converted into a new space to achieve dimensional reduction of data features. However, in our study, PCA is used to improve the estimation performance, rather than dimension reduction, which is the same PCA application described in [15].

2.4.3. Random Forest with Principal Component Analysis

The structural diagram of the RFPCA method used for the knee estimation from the sEMG is shown in Figure 5. In this figure, the blue lines represent the training process and the red lines describe the studying course. The time-domain amplitudes were extracted from 5 sEMG channels. After the process and PCA, one part of the data would be used for training to build the estimation model and the rest would be used for testing.

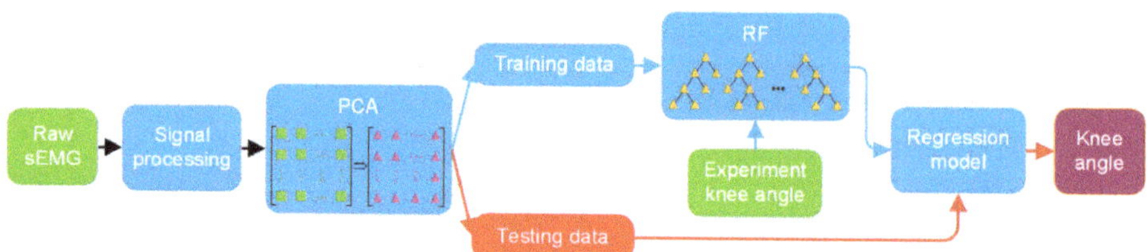

Figure 5. The structural diagram of the random forest with principal component analysis (RFPCA) used for knee estimation from sEMG.

In our work, the processed sEMG $\boldsymbol{X}_t$ forms the original data set $\boldsymbol{X}_{emg}$, and $\boldsymbol{X}_{emg}$ which is then transformed through a matrix $\boldsymbol{P}$ by PCA. Next, the input $\boldsymbol{X}$ of the following RF is obtained, which is described as follows:

$$\boldsymbol{X} = \boldsymbol{P}\boldsymbol{X}_{emg}. \tag{3}$$

with the combination of $\boldsymbol{X}$ and the knee angle y_{knee}, the data set after PCA is derived as $(\boldsymbol{X}, y_{knee})$. These data are divided into training set and testing set.

When the training sample $\boldsymbol{T}_r = (\boldsymbol{X}_{tr}, \boldsymbol{Y}_{tr})$ is given, the goal is to use the $\boldsymbol{T}_r$ to establish a estimation model $E(\boldsymbol{X})$ and apply it to estimate the new knee angle from sEMG. The RF consists of a collection of N randomized regression trees $r(\boldsymbol{X}, v_i)$, where v_i ($i = 1, 2, \ldots, N$) are the independent random variables. They are used to resample the training set and select the successive direction for splitting [19]. The estimation model $E(\boldsymbol{X})$ is an average of the regression trees in the RF, expressed as:

$$E(\boldsymbol{X}) = \frac{\sum_{i=1}^{N} r(\boldsymbol{X}, v_i)}{N}, \tag{4}$$

When the testing data set $\boldsymbol{X}_{test}$ is used to validate the estimation performance, the estimated knee angle can be calculated as follows:

$$\hat{y}_{knee} = E(\boldsymbol{X}_{test}; v_1, v_2, \cdots, v_N, \boldsymbol{T}_r) = \frac{1}{N}\sum_{i=1}^{N} r(\boldsymbol{X}_{test}; v_i, \boldsymbol{T}_r). \tag{5}$$

2.5. Arguments Selection

For comparison with RFPCA in this study, another method, namely BPPCA, was also presented. This method is similar to RFPCA in Section 2.4.3, with a substitution of a conventional BP for RF. For different methods, there are different arguments regarding the effect they have on the performance of the estimation. In a model of RF, there are three main parameters to be determined, namely the number of trees in the forest (N), the number of features of the input (N_f) and the minimum size of the terminal nodes (N_m). The parameters for the BP are the number of epochs (N_e), the number of hidden layer nodes (N_h), the learning rate (L_r) and the learning goal (L_g).

During the test of the ML methods, the results showed that the performance of motion estimation was not sensitive to the difference of individuals using the same parameters, which was similar to the results found in [19]. Thus, it meant that universal values could be chosen from these parameters. Referring to the parameter settings in [19], and to reach a compromise between estimation accuracy and computational time, the parameters for our methods were set as recorded in Table 2.

Table 2. The values of parameters of the machine learning (ML) methods.

Argument	N	N_f	N_m	N_e	N_h	L_r	L_g
Value	50	5	5	100	40	0.1	0.00001

To study the historic effect of previous sEMG, the data from one trail with 4 GCs from each subject would be introduced. A total of 75% of the data was used for training and the rest was used for the testing set.

For the study of the sample size, as noted, 61 GDs were utilized for the estimation work of one subject, and the last GD, with approximately 128 samples, was always the testing set. Hereby, parameter S_s in the interval of [1, 60] was defined to represent the sample size. For example, when $S_s = 1$, the 60th GD would be the training set, when $S_s = 2$, the 60th and the 59th GDs were the training set, and by that logic, when S_s was 60, all of the 60 GDs would be used for training. The sample size was nearly proportional to the S_s.

2.6. Evaluation Metric

The performance of the estimation model was mainly evaluated by comparing the root mean square error (R) between the estimated value $\hat{y}_{knee}$ and the experiment value (EV) y_{knee} defined as follows:

$$R = \sqrt{\frac{1}{S}\sum_{i=1}^{S}\left(\hat{y}_{knee,i} - y_{knee,i}\right)^2}, \tag{6}$$

where S is the number of test samples. In this work, R is an indicator for the judgement of the estimation models, and the accuracy of the estimation model increases with a decrease of R.

3. Results

In this section, different kinds of estimation methods for the estimation of the knee from the sEMG signals were compared to investigate whether the sample size and the previous sEMG signals would affect the performance of the estimation. All of the data processing was done on the software of MATLAB R2016b. In order to avoid the influence of the processor on the computation time, all the execution time in our study is relative.

3.1. The Results of Different Sample Size

Different sample sizes from 1 GD to 60 GDs were utilized for RFPCA and BPPCA to predict the knee joint angle.

Partial results of Subject 1 for estimation are shown in Figure 6. It can be seen that, as S_s increases, the estimations of BPPCA and RFPCA are close to the EV. However, the estimation results of RFPCA are always closer to the EV and more robust than BPPCA.

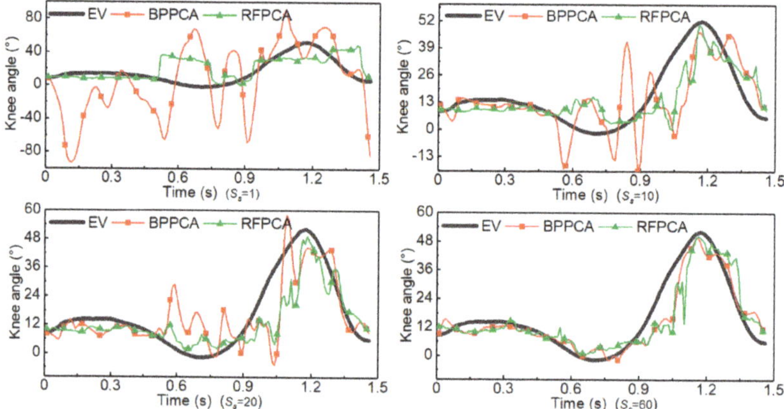

Figure 6. Partial results for estimation with different sample sizes of Subject 1.

The relative execution time (the execution time of each calculation processed against the longest computation among all of the sample size results), and the evaluation metric for the estimation with different methods and different subjects, are depicted in Figure 7. As the sample size increases, so does the execution time. Compared to the increase in time using BPPCA, RFPCA appears more efficient for the training, as it is less time consuming. Although the value of R decreases and the error decreases with the larger sample size from 1 to 10 GDs, both of the methods have no significant changes after $S_s = 20$. When $S_s \geq 30$, the R of BPPCA is slightly smaller than RFPCA, with a longer execution time.

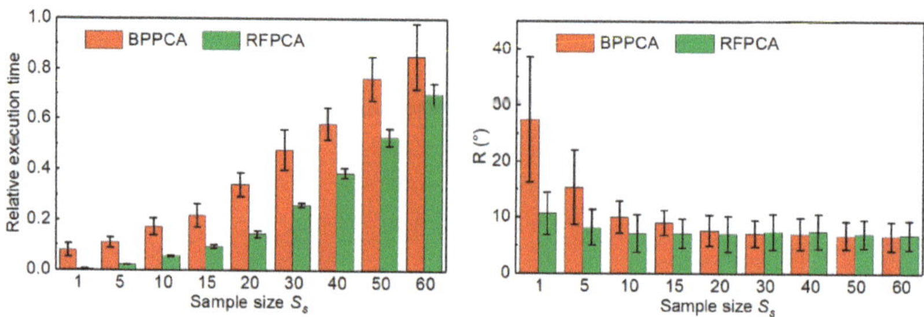

Figure 7. The relative execution time and R for the estimation with different S_s.

3.2. The Results of Different Previous sEMG Input

In the process of training, different previous sEMG signals were used, and the result of relative execution time and the evaluation metric are shown in Figure 8. As the attributes of input $\mathbf{X}_t$ increase, the execution time of the different methods also increases, and this is more pronounced for BPPCA. RFPCA is much smaller, which is conducive to the application of myoelectric control. For the results of R, the estimation errors of different methods are various. The BPPCA results are much larger than RFPCA, and the method is insensitive to the input dimension when n is larger than approximately 7. The results of RFPCA seem to increase when n is bigger than 2. The results of R also show that BPPCA has larger standard deviations than RFPCA.

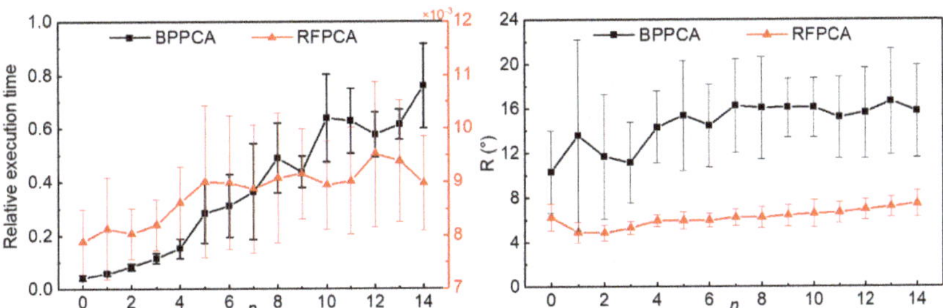

Figure 8. The relative execution time and R for the estimation with different previous input.

The estimation results of all six subjects using various methods when $n = 2$ are shown in Figure 9. From the figure, it is clear that the EV tracking errors of RFPCA are much smaller than BPPCA.

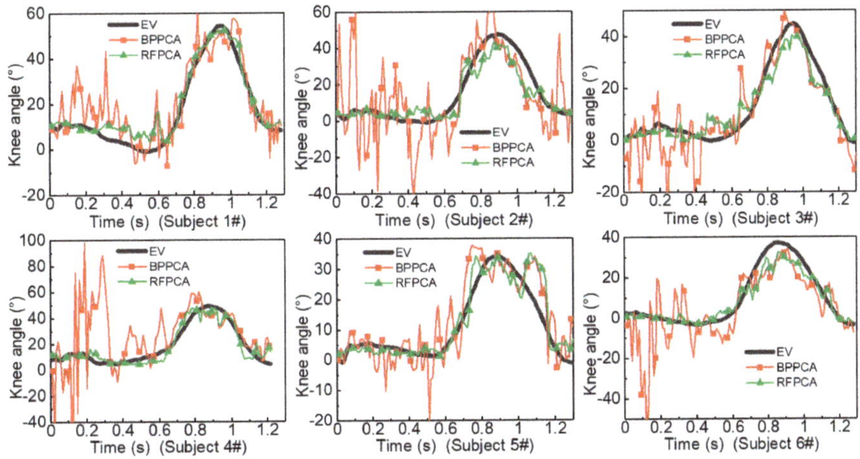

Figure 9. The estimation results of different models from different subjects when $n = 2$.

Furthermore, the evaluation metric R of different models for different subjects when $n = 2$ is depicted in Figure 10. The R of RFPCA is almost 5°, while the BPPCA has a poor prediction ability of the knee angle estimation, showing a large variation between different subjects, with errors ranging between 7° to 25°.

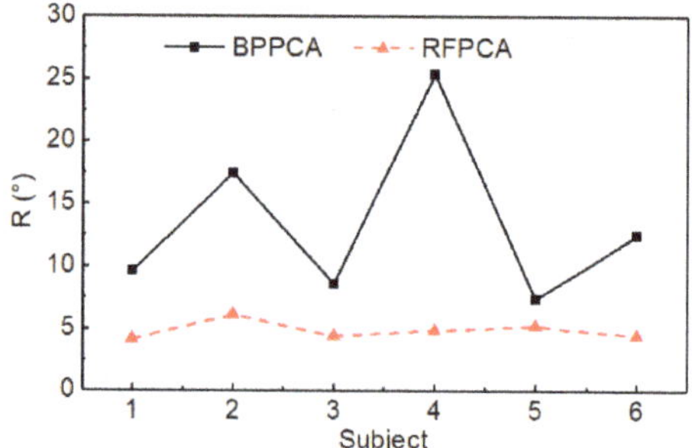

Figure 10. The evaluation metric of different models for different subjects when $n = 2$.

4. Discussion

In this paper, a novel estimation method of RFPCA was proposed to study the relationship between sEMG and knee movement. Compared with the results of BPPCA, the RFPCA performs better, both in terms of the root mean square error and the execution time. All of the estimation results using RFPCA are also generally in line with the EV. These results may be due to the strong regression ability of RF, which generates an internal unbiased estimate of the generalization error as the forest building progresses. PCA is able to generate a better input for RF from the original data, which also promotes the accuracy of the results of estimation.

As seen in Figure 7, with the increasing of input samples, R starts to decrease and eventually stabilizes, which means that the prediction accuracy increases at first, and then does not change significantly for both methods. As known, walking is a regular movement, the kinematic parameter of the gait is a cyclic process and the sEMG also appears as a periodic signal in different GCs. Thus, we believe that when the sample size increases to a certain value, the differences between the samples decrease, so that the prediction results show little change. However, the larger the data, the longer the time. With acceptable accuracy, choosing a better sample size can effectively reduce the time consumption caused by large samples,, and this would contribute to the efficiency of online control using sEMG.

The historic effect of the input has a positive influence on motion estimation, according to the authors of [19]. In this work, as the input size increases, there is a tendency for the R of both RFPCA and BPPCA to increase, as shown in Figure 8, and the previous signals seem to have little to no effect on the estimation after more dimensional data is involved in the calculation. Generally, in the high-dimensional input case, the problem of sparse data samples and the difficulty of distance calculation are a common and serious obstacle for all machine learning methods, which is called the "curse of dimensionality". Thus, except for the PCA used in our work, further study of the input dimension of sEMG needs to be considered. Multichannel sensors of sEMG detecting are also worthy of research.

As seen in Figure 9, in terms of the physical and mental quality of the test subjects, the prediction results of different subjects also vary. In addition, the skin preparations of the subjects vary from one subject to another, which can also contribute to detection errors of in the raw sEMG. Moreover, different times for the experiment and other environmental factors will also cause diversity in the outcome when the raw sEMG is being collected. However, both Figures 9 and 10 show that the results of the BPPCA are more unstable for all participants, relatively, and that the results of the RFPCA have estimations similar to the EV. That is, RFPCA has a better error tolerance and is not adversely affected by variations in test subjects.

The root mean square error of RFPCA was small ($R \approx 5°$) in the previous sEMG study when $n = 2$, as shown in Figure 10, and this result seems to be large in comparison to the motion angle of the knee. However, we believe that this is acceptable for the exoskeleton control, since the control precision of the exoskeleton joint is not in high demand as a machine tool, and the pilot of the exoskeleton is able to tolerate the small differences of several degrees due to the flexibility of the human body. Also, the estimation model using RFPCA has some hyperparameters in the structure of the forest. Therefore, a better parameter selection may lead to a more accurate estimation model of RFPCA, and this would also help the application of RFPCA for estimation of joint movement in myoelectric control. Furthermore, while the RFPCA performed better in our study, the BPPCA may have advantages over the RFPCA with a better parameter choice.

Since the aim of this work is using ML to build an estimation model which can be adjusted to suit the subject himself, the training and testing data are from the same subject for each validation, and a similar method can be found in [30]. The results of the subjects in this work using RFPCA was acceptable. However, sEMG is unstable, as mentioned above, and the RFPCA method may be unfeasible for subjects who are middle-aged and elderly, since the subjects in our study were under 30 years of age, and the data from different people for training and testing may be an issue through the method of RFPCA. Thus, more people will be invited to participate in the study to further validate the proposed method.

5. Conclusions

In this study, a new method of combining RF and PCA was utilized to establish an estimation model from sEMG to the knee. Better than BPPCA, the RFPCA method is able to predict the knee angle at a low root mean square error about 5° and the execution time is several times smaller, and these results indicate that RFPCA is suitable for knee movement estimation with high accuracy and requiring minimal time. Both the sample size and the input dimension of the RFPCA were investigated. RFPCA is insensitive to the estimation accuracy when the sample size increases to a certain value, but requires more time. Also, as the previous sEMG increases, the accuracy of the RFPCA first increases and then decreases, and the work would be beneficial to the estimation model construction using ML methods. Moreover, the results of the RFPCA are more stable in comparison to the BPPCA in terms of sample size and previous sEMG input, and RFPCA is also robust in terms of differences among test subjects. All in all, the estimation of RFPCA performs well in the estimation from sEMG to knee movement in this work, which is conductive to motion analysis and exoskeleton control.

Future work of improving the estimation accuracy will be done by extracting more signal features or fusing other physical sensors, as well as testing more people and other daily activities such as different speeds of walking, and ascending or descending stairs. Furthermore, a wider application for RFPCA may be utilized for estimation using different sEMG data from various people to find a general relationship between sEMG and joint motion. Finally, the RFPCA method will be applied to estimate human motion for exoskeleton control.

Author Contributions: Conceptualization, Z.L., X.G. and C.X.; methodology, Z.L. and X.G.; software, Z.L.; validation, Z.L., X.G. and C.X.; formal analysis, Z.L. and X.G.; investigation, Z.L. and K.Z.; resources, Z.L. and K.Z.; data curation, Z.L., X.G. and C.X.; writing—original draft preparation, all the authors; writing—review and editing, all the authors; visualization, Z.L. and X.G.; supervision, X.G. and C.X.; project administration, X.G.; funding acquisition, X.G. All authors have read and agreed to the published version of the manuscript.

Funding: This research was funded by NATIONAL DEFENSE BASIC SCIENTIFIC RESEARCH PROGRAM OF CHINA, grant number B1020132012.

Acknowledgments: Thanks to the help of the participants in the experiment and the support of Laboratory for Individual Equipment Technology at School of Mechanical Engineering, Nanjing University of Science and Technology. We are also grateful for the assistance of the editors and reviewers.

Conflicts of Interest: The authors declare no conflict of interest.

References

1. Fan, Y.; Yin, Y. Active and Progressive Exoskeleton Rehabilitation Using Multisource Information Fusion from EMG and Force-Position EPP. *IEEE Trans. Biomed. Eng.* **2013**, *60*, 3314–3321. [CrossRef]
2. Chen, C.; Du, Z.; He, L.; Wang, J.; Wu, D.; Dong, W. Active Disturbance Rejection with Fast Terminal Sliding Mode Control for a Lower Limb Exoskeleton in Swing Phase. *IEEE Access* **2019**, *7*, 72343–72357. [CrossRef]
3. Chae, J.; Jin, Y.; Sung, Y.; Cho, K. Genetic Algorithm-Based Motion Estimation Method using Orientations and EMGs for Robot Controls. *Sensors* **2018**, *18*, 183. [CrossRef]
4. Kang, I.; Kunapuli, P.; Hsu, H.; Young, A.J. Electromyography (EMG) Signal Contributions in Speed and Slope Estimation Using Robotic Exoskeletons. In Proceedings of the 16th International Conference on Rehabilitation Robotics, Toronto, ON, Canada, 24–28 June 2019; pp. 548–553. [CrossRef]
5. Moon, D.H.; Kim, D.; Hong, Y.D. Intention Detection Using Physical Sensors and Electromyogram for a Single Leg Knee Exoskeleton. *Sensors* **2019**, *19*, 4447. [CrossRef]
6. Kartsch, V.; Benatti, S.; Mancini, M.; Magno, M.; Benini, L. Smart Wearable Wristband for EMG based Gesture Recognition Powered by Solar Energy Harvester. In Proceedings of the IEEE International Symposium on Circuits and Systems, Florence, Italy, 27–30 May 2018; pp. 1–5. [CrossRef]
7. Resnik, L.; Huang, H.; Winslow, A.; Crouch, D.L.; Zhang, F.; Wolk, N. Evaluation of EMG pattern recognition for upper limb prosthesis control: A case study in comparison with direct myoelectric control. *J. Neuroeng. Rehabil.* **2018**, *15*, 23:1–23:13. [CrossRef]
8. Jang, G.; Kim, J.; Lee, S.; Choi, Y. EMG-based continuous control scheme with simple classifier for electric-powered wheelchair. *IEEE Trans. Ind. Electron.* **2016**, *63*, 3695–3705. [CrossRef]
9. Chen, J.; Zhang, X.; Gu, L.; Nelson, C. Estimating Muscle Forces and Knee Joint Torque Using Surface Electromyography: A Musculoskeletal Biomechanical Model. *J. Mech. Med. Biol.* **2017**, *17*, 3:1–3:21. [CrossRef]
10. Tagliapietra, L.; Vivian, M.; Sartori, M.; Farina, D.; Reggiani, M. Estimating EMG signals to drive neuromusculoskeletal models in cyclic rehabilitation movements. In Proceedings of the 37th Annual International Conference of the IEEE Engineering in Medicine and Biology Society (EMBC), Milan, Italy, 25–29 August 2015; pp. 3611–3614. [CrossRef]
11. Zhuang, Y.; Yao, S.; Ma, C.; Song, R. Admittance Control Based on EMG-Driven Musculoskeletal Model Improves the Human–Robot Synchronization. *IEEE Trans. Ind. Inf.* **2018**, *15*, 1211–1218. [CrossRef]
12. Anwar, T.; Al-Jumaily, A.; Watsford, M. Estimation of Torque Based on EMG using ANFIS. *Procedia Comput. Sci.* **2017**, *105*, 197–202. [CrossRef]
13. Gui, K.; Liu, H.; Zhang, D. A Practical and Adaptive Method to Achieve EMG-Based Torque Estimation for a Robotic Exoskeleton. *IEEE-ASME Trans. Mech.* **2019**, *24*, 483–494. [CrossRef]
14. Bi, L.; Feleke, A.G.; Guan, C. A review on EMG-based motor intention prediction of continuous human upper limb motion for human-robot collaboration. *Biomed. Signal Process.* **2019**, *51*, 113–127. [CrossRef]
15. Toledo-Pérez, D.; Martínez-Prado, M.; Gómez-Loenzo, R.; Paredes-García, W.; Rodríguez-Reséndiz, J. A Study of Movement Classification of the Lower Limb Based on up to 4-EMG Channels. *Electronics* **2019**, *8*, 259. [CrossRef]
16. Morbidoni, C.; Cucchiarelli, A.; Fioretti, S.; Di Nardo, F. A Deep Learning Approach to EMG-Based Classification of Gait Phases during Level Ground Walking. *Electronics* **2019**, *8*, 894. [CrossRef]
17. Nazmi, N.; Abdul Rahman, M.A.; Yamamoto, S.; Ahmad, S.A. Walking gait event detection based on electromyography signals using artificial neural network. *Biomed. Signal Process.* **2019**, *47*, 334–343. [CrossRef]
18. Bao, T.; Zaidi, A.; Xie, S.; Zhang, Z. Surface-EMG based Wrist Kinematics Estimation using Convolutional Neural Network. In Proceedings of the 16th International Conference on Wearable and Implantable Body Sensor Networks (BSN), Chicago, IL, USA, 19–22 May 2019; pp. 1–4. [CrossRef]
19. Xiao, F.; Wang, Y.; Gao, Y.; Zhu, Y.; Zhao, J. Continuous estimation of joint angle from electromyography using multiple time-delayed features and random forests. *Biomed. Signal Process.* **2018**, *39*, 303–311. [CrossRef]
20. Lei, Z. An upper limb movement estimation from electromyography by using BP neural network. *Biomed. Signal Process.* **2019**, *49*, 434–439. [CrossRef]
21. Huang, Y.; He, Z.; Liu, Y.; Yang, R.; Zhang, X.; Cheng, G.; Yi, J.; Ferreira, J.P.; Liu, T. Real-Time Intended Knee Joint Motion Prediction by Deep-Recurrent Neural Networks. *IEEE Sens. J.* **2019**, *19*, 11503–11509. [CrossRef]
22. Hermens, H.J.; Freriks, B. Welcome to SENIAM. Available online: http://www.seniam.org/ (accessed on 5 June 2019).

23. Konrad, P. *The ABC of EMG: A Practical Introduction to Kinesiological Electromyography*; Noraxon USA Inc.: Scottsdale, AZ, USA, 2006; ISBN 0-9771622-1-4.
24. Breiman, L. Random forests. *Mach. Learn.* **2001**, *45*, 5–32. [CrossRef]
25. Liaw, A.; Wiener, M. Classification and Regression by RandomForest. *R News* **2002**, *2*, 18–22.
26. Breiman, L.; Cutler, A. Random Forests. Available online: https://www.stat.berkeley.edu/~breiman/RandomForests/cc_home.htm (accessed on 21 September 2019).
27. Pearson, K. LIII. On lines and planes of closest fit to systems of points in space. *Lond. Edinburgh Dublin Philos. Mag. J. Sci.* **1901**, *2*, 559–572. [CrossRef]
28. Hotelling, H. Analysis of a complex of statistical variables into principal components. *J. Educ. Psychol.* **1933**, *24*, 417–441. [CrossRef]
29. Goodfellow, I.; Bengio, Y.; Courville, A. *Deep Learning*; MIT Press: Cambridge, MA, USA, 2016; Volume 2, pp. 45–46.
30. Morbidoni, C.; Principi, L.; Mascia, G.; Strazza, A.; Verdini, F.; Cucchiarelli, A.; Di Nardo, F. Gait Phase Classification from Surface EMG Signals Using Neural Networks. *IFMBE Proc. Springer Cham.* **2020**, *76*, 75–82. [CrossRef]

Article

A Wireless Body Sensor Network for Clinical Assessment of the Flexion-Relaxation Phenomenon

Michele Paoletti [1], Alberto Belli [1], Lorenzo Palma [1], Massimo Vallasciani [2] and Paola Pierleoni [1,*]

1 Department of Information Engineering (DII), Università Politecnica delle Marche, 60131 Ancona, Italy; m.paoletti@pm.univpm.it (M.P.); a.belli@univpm.it (A.B.); l.palma@pm.univpm.it (L.P.)

2 Istituto di Riabilitazione Santo Stefano, 62018 Porto Potenza Picena, Italy; massimo.vallasciani@sstefano.it

* Correspondence: p.pierleoni@univpm.it; Tel.: +39-0712204847

Received: 27 May 2020; Accepted: 20 June 2020; Published: 24 June 2020

Abstract: An accurate clinical assessment of the flexion-relaxation phenomenon on back muscles requires objective tools for the analysis of surface electromyography signals correlated with the real movement performed by the subject during the flexion-relaxation test. This paper deepens the evaluation of the flexion-relaxation phenomenon using a wireless body sensor network consisting of sEMG sensors in association with a wearable device that integrates accelerometer, gyroscope, and magnetometer. The raw data collected from the sensors during the flexion relaxation test are processed by an algorithm able to identify the phases of which the test is composed, provide an evaluation of the myoelectric activity and automatically detect the phenomenon presence/absence. The developed algorithm was used to process the data collected in an acquisition campaign conducted to evaluate the flexion-relaxation phenomenon on back muscles of subjects with and without Low Back Pain. The results have shown that the proposed method is significant for myoelectric silence detection and for clinical assessment of electromyography activity patterns.

Keywords: flexion-relaxation phenomenon; surface electromyography; wearable device; WBSN; automatic detection of the FRP

---

1. Introduction

The Flexion-Relaxation Phenomenon (FRP) term was adopted in 1955 by Floyd and Silver analysing the erector spinae muscles [1]. It consists of a back muscle electrical activity silence which typically occurs during the trunk full flexion. This effect is believed to be the result of ligaments activity and other passive elements of the spine that absorb the load of muscles. The erector spinae muscles (extensors of the trunk) contract when the trunk is flexed from the upright position acting as gravity antagonists. Floyd and Silver observed that myoelectric quiescence was caused by a reflex due to stretching in which the load torque of the upper body was transferred from the active to the passive spinal elements. It was also shown that, although the surface muscle activity was electrically very reduced, muscles continued to provide support through the stretching of the passive elements [2], and some of the deep muscles remaining electrically active in load support [3]. In the literature, it is known that the pain interferes with both afferent and efferent aspects of neuromuscular control [4–6]. Generally, in healthy subjects without Low Back Pain (LBP) stories, the FRP is statistically present, while in LBP patients the phenomenon is frequently absent. In order to evaluate if the subject has normal neuromuscular patterns amongst the various physiological indicators of LBP, the FRP has been one of the most studied surface electromyographic responses in the literature [1,3,7–9]. A lack of FRP was significant in pathological patients with pain, perceived disability, and re-injury fear. Furthermore, cases of healthy subjects without FRP, and LBP subjects with FRP (typically when the pain is chronic) have been reported, but they were less frequent [10]. Sihvonen has reported that the

FRP absence was more easily observed in subjects with LBP presence than patients without pain during the test [11]. Since the FRP absence is often used as an indicator for low back dysfunctions [12,13], several studies have used different methods to quantify the myoelectric activity, and discriminate the subjects, knowing a priori the health conditions [10,14–16]. The Flexion Relaxation Ratio (FRR) is one of these methods used to quantify FRP level and try to discriminate healthy subjects from LBP patients [11].

Some research works proposed in the literature have investigated the FRP phenomenon [17,18] and the evaluation methods to identify its presence/absence in healthy subjects [19,20]. An accurate method for the FRP analysis is the Visual Inspection Method (VIS), consisting of visually identify the phenomenon presence or absence by a subjective analysis of the processed sEMG signal, on the selected muscles [20]. This approach requires experience, it cannot be used by non-expert examiners and it causes a strong waste of time. However, thanks to its reliability, the VIS method was used to compare the algorithm performances to detect the FRP in healthy subjects [20]. Because muscular activities appear during the relaxation phase, in the case of LBP subjects, the automatic evaluation of the relaxation phase limits is a non-trivial problem for this kind of patients compared to healthy subjects [21] and automatic methods are needed to accurately quantify first a parameter associated with the FRP level (for example the FRR) and then identify FRP presence/absence. Alison et al. have used two types of VIS: VIS1 based on the raw EMG data and VIS2 based on the linear envelope of the EMG data [20]. However, the visual inspection only based on the sEMG signal is possible only when the patterns are recognizable through a visual analysis (typically, as in the case of Alison, when there are completely healthy subjects with regular patterns).

In the literature, several studies have proposed systems for the evaluation of FRP only on healthy subjects. In particular, Ritvanen et al. [22] have used one sEMG system while Alison Schinkel-Ivy et al. [18] have added a motion capture system to it. Sihvonen et al. [23] have proposed a similar system based on a wired Body Sensor Network. In this study, we wish to evaluate the performance of a Wireless Body Sensor Network (WBSN) able to analyse, quantitatively and objectively, the surface electromyography of the low back muscles (longissimus and multifidus) and automatically detect FRP presence/absence in both healthy subjects and patients with LBP. We focused attention on the quantification and evaluation of FRP rather than the discrimination in healthy and LBP subjects starting from the FRP obtained (without knowing the health conditions), which represents the next step of the analysis.

Generally, the FRP phenomenon is evaluated by observing the surface electrical activity of the spinal extensor muscles during a motion task where the subject reaches the maximal trunk flexion and returns to an upright position. This motion task, also known as flexion-relaxation test (or forward bend test), is mainly composed of four phases: standing, flexion, full flexion, and extension. In this test, starting from the upright position (Standing phase), the subject bends forward (Flexion phase). Once the bending of the torso reaches the maximum bending value of the trunk (Full-Flexion phase) naturally without straining the back, the subject returns to the initial position (Extension phase). During the execution, the lumbar spine is exploited in the first 50° of flexion phase, while in the remaining degrees the flexion occurs through the rotation of the pelvis [24]. In the extension phase instead inverse happens: rotation of the pelvis followed then by a lumbar spine extension. FRP statistically occurs in healthy subjects when the spine is about at 90°, compared to the standing position, which is in the full flexion phase [25,26]. Therefore, an accurate FRP clinical assessment on back muscles requires a method for the surface electromyography signals analysis, especially in the full flexion phase. It is necessary, in order to identify each phase of the flexion-relaxation test, to process the data acquired by a system able to estimate the inclination of the subject during the bending movement. Motion analysis systems [27–29] and electronic goniometers [30] have been proposed for inclination detection during the flexion-relaxation movement. However, video-based analysis is time-consuming and limited to a given space under observation, while electronic goniometers are bulky and, therefore, they can hinder the natural movement. In order to overcome these limitations, we have used a system based on non-obtrusive wearable and portable devices able to estimate the subject's inclination [31–33]. Such devices, compared to the electronic goniometers, have achieved an average error less of 4° for angle estimation and movement analysis [34].

Systems based on wearable devices have obtained similar performances when they were compared to optoelectronic systems (they represent the gold standard for motion analysis). As demonstrated in the literature, a comparative study has been proposed to validate a system composed of two wearable devices for lumbar inclination detection and it has shown an average error less of 2° compared to an optoelectronic system [35]. Despite the use of wearable devices was proposed and validated in gait analysis [36,37], there isn't currently a common agreement on which is the most appropriate approach to estimate the subject's inclination [38,39] and identify the different phases in the flexion-relaxation test to observe the electromyography signals and facilitate the FRP identification (with VIS and FRR methods).

In this paper, we propose a WBSN for the clinical assessment of the FRP during the flexion-relaxation test. It is composed of two separate systems: non-invasive wireless surface electromyography (sEMG) sensors in association with an inertial wearable device. With the algorithm developed by us, through a separate signal processing, subsequent synchronization, and overlap, we have obtained a single integrated network of sensors. Recent studies have proposed a system for assessment of the FRP even in LBP subjects. In particular, Ducina et al. have used a 12-camera motion analysis system to determine the inclination [40]. Other studies have started to use wearable wireless inertial systems to study the FRP [41,42]. However, these studies have not proposed automatic algorithms for detecting FRP or at least have not provided the results obtained with them. In this study, starting from the data acquired by the WBSN, an algorithm based on the FRR method was implemented to automatically detect the FRP and the obtained results have been compared with that deriving from the visual inspection. The developed software was tested using a dataset composed of healthy subjects (also called controls) and LBP subjects (also called patients) [43].

2. Materials and Methods

Given the lack of a standard procedure for clinical FRP assessment on back muscles, in this work, we have proposed a WBSN to automatically analyse the multichannel surface-electromyography signals together with the real movement performed by the subject. It consists of a wireless surface electromyography system composed of four sEMG sensors and one wearable device with triaxial accelerometer, gyroscope, and magnetometer sensors embedded. As shown in Figure 1, the wearable device was positioned on the first lumbar vertebra and the electrodes of the 4 sEMG sensors were placed along the fibers [44] of the longissimus left, longissimus right, multifidus left, multifidus right muscles, respectively. Longissimus muscles are part of the erector muscles of the vertebral column together with the iliocostalis and spinal muscles. The myoelectric signals of the longissimus muscles were acquired from two pairs of electrodes (LSX and LDX) positioned two fingers apart in a lateral direction from the spinous process L1 [45]. Multifidus muscles are part of the deep muscles of the trunk as they are in close contact with the spine. The myoelectric signals of the multifidus muscles were acquired from two pairs of electrodes (MSX and MDX) positioned on the line connecting the caudal tip of the posterior superior iliac spine to the space between L1 and L2, at the level of the spinous process of L5, 2–3 cm from the medial line [45]. The application surface was cleaned, before applying electrodes on the skin, with an alcohol swab, and the appropriate professional paste was applied to the electrodes in order to reduce the effects of the resistance provided by the skin and impurities [46,47].

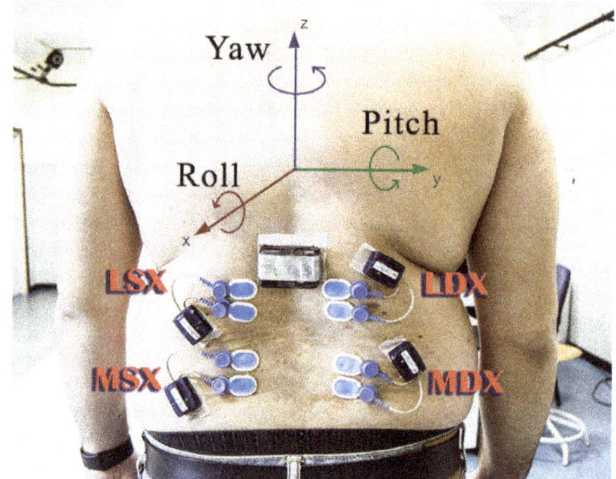

Figure 1. Positioning of the sEMG sensors and the wearable device on the subject under analysis. The electrodes were positioned following European recommendations for surface electromyography [45].

In order to identify muscles activity, the sEMG signals were acquired with a sampling frequency of 2000 Hz. The acquired sEMG signals were the following:

- Electromyography signal on left longissimus channel (LSX);
- Electromyography signal on right longissimus channel (LDX);
- Electromyography signal on left multifidus channel (MSX);
- Electromyography signal on right multifidus channel (MDX).

The wearable device signal, to estimate the inclination of the subject, was acquired with a sampling frequency of 128 Hz. The data collected by the wearable device were the following:

- Acceleration measured by the accelerometer (ACC);
- Angular velocity measured by the gyroscope (GYR);
- Magnetic field measured by the magnetometer (MAG).

The electromyography signals obtained by each sEMG sensor and the data collected by the wearable device were sent to a personal computer that acted as a central processing unit. The data acquired during the flexion-relaxation test, by the subjects under analysis, were processed and stored [12]. Before starting the forward bend test, the subject was placed with the arms on the side with the feet to the width of the shoulders, standing upright with the gaze straight and fixed on one point in order to avoid any artifact due to the alteration of the head position. During the flexion-relaxation test, the subject wore the proposed WBSN and he repeated 4 times a motion trial in which he was asked to naturally reach a bend angle about 90° without straining the lumbar region. One complete movement was called "cycle" and it was repeated 4 times (a compromise that allowed to have a good number of repetitions without exaggerating and stressing too much the muscles involved). As shown in Figure 2, each cycle consists of 4 phases listed below:

1. Standing—The subject keeps the standing position for about 4 s;
2. Flexion—The subject bend forward in order to naturally reach the full flexion position;
3. Full flexion—The subject keeps the full flexion position for about 4 s;
4. Extension—The subject return to standing position.

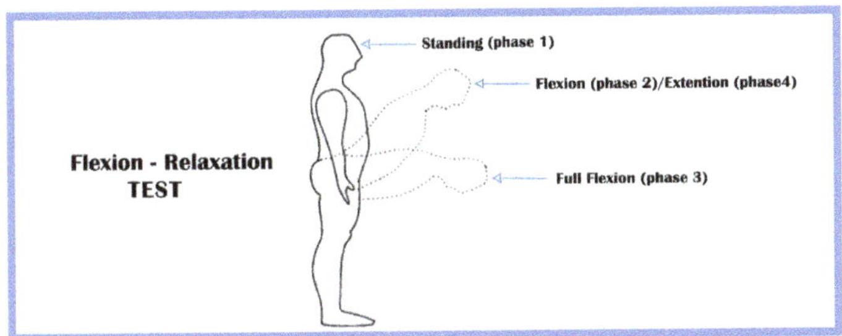

Figure 2. Representation of the movement performed by the subject during the flexion relaxation test.

According to McGorry et al. [12], the FRP may vary with changes in execution speed, prolonged static flexion, rest time, external load application. So, we must try to keep these parameters as constant as possible in order to make the results comparable over time.

The proposed WBSN was used in an extensive acquisition campaign conducted in collaboration with Santo Stefano Rehabilitation Institute (Porto Potenza Picena, Italy). The purpose was also to collect data from healthy and LBP subjects, to investigate the relationship between FRP and physical conditions. Before starting the acquisition campaign, demographic data, and patient history of each subject were registered to have a report about perceived pain and disability conditions. NRS-11 scale was used to identify the perceived pain during, before, and after the flexion-relaxation test execution [43,48]. Another patient condition measure (disability) was evaluated with the backill questionnaire [49] that assesses the ability to make or not a series of activities [43]. Procedures and experimental design of the acquisition campaign have been described in our previous study together with the complete dataset used to evaluate the FRP detection performances of the developed algorithm [43]. The dataset includes information and signals acquired on a total of 25 subjects submitted to the flexion-relaxation test, using the proposed WBSN.

3. Algorithm for FRP Clinical Assessment

An accurate FRP clinical assessment requires the analysis of the electromyography signals on back muscles and the measurement of the subject's inclination during the flexion-relaxation test. The WBSN automatically provides a clinical evaluation of the FRP through the algorithm proposed in this paper and described in the block diagram of Figure 3.

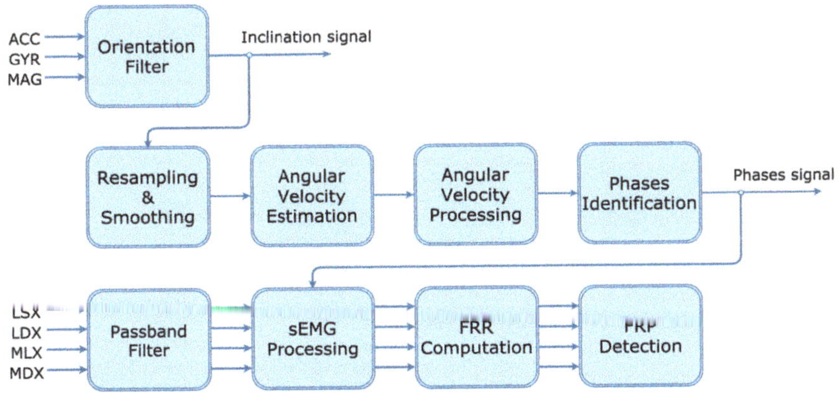

Figure 3. Block diagram of the proposed algorithm for clinical assessment of the FRP

Starting from the raw data acquired by the wearable device, an orientation filter [50,51] was implemented to provide the orientation of the subject's trunk in terms of Yaw, Pitch, and Roll angles [52].

As shown in Figure 1, Yaw, Pitch, and Roll angles describe the rotations around the Z, Y and X axes, respectively. Such representation describes the actual orientation of the subject's trunk starting from a fixed initial frame and the pitch angle identifies the inclination signal used to be associated with each electromyography signal acquired from the back muscles. Processing the inclination data it is possible to obtain a signal, called "phases signal", to define the onset/end phase of the motion, on the flexion-extension test. The main steps used by the proposed algorithm to compute the phases signal are described in Figure 3, and an example of the inclination signal processing is shown in Figure 4. Due to the different sampling frequency compared to the sEMG sensors system, the inclination signal was resampled to 2 kHz using a linear interpolation function. This increase in frequency of the inclination signal, albeit fictitious through oversampling, is not significant due the sampling frequency used is high enough to fully describe the specific movement. Subsequently, the resampled inclination signal was processed using a moving average filter with a span equal to 100 (orange signal in Figure 4). The derivative of the previous signal respect the time ($d\theta/dt$) was made in order to compute the angular velocity and its absolute value (blue signal in Figure 4). To obtain a signal that was zero when there were variations and changes during the static phases the reciprocal of the absolute value of angular velocity was computed and subsequently filtered using a 3th-order one-dimensional median filter and a time moving average filter with a span equal to 100 (violet signal in Figure 4). This signal, called processed angular velocity, was compared with an empirical threshold level equal to 0.09, to produce the phases signal (green signal in Figure 4). This threshold value was empirically obtained by carrying out a series of tests. In particular, several acquisitions on healthy subjects with normal sEMG patterns were analysed by superimposing the "Inclination signal" on the "Processed angular velocity signal". The chosen threshold represented the best value able to identify the various phases since the sEMG patterns with that value coincided perfectly with the duration of the phases. When the processed angular velocity was greater than the threshold level, the phases signal was set to a high value to identify standing and full-flexion phases. Otherwise, when the processed angular velocity was under the threshold level the phases signal was set to a low value to identify flexion and extension phases. Thus, the proposed algorithm is able to identify: phase onset/end, cycle onset/end, flexion-extension test onset/end, discarding the samples of the signals that exceed the last phase of the last cycle.

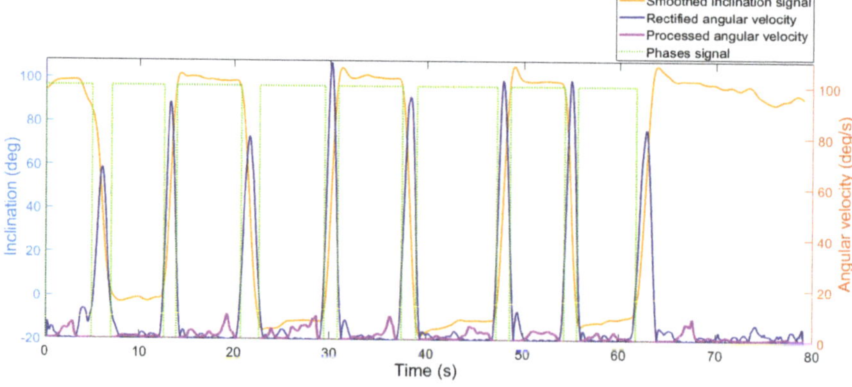

Figure 4. Signal processing of the inclination signal in order to obtain the "phases signal" which automatically defines the phases and cycles during a flexion-relaxation test.

In order to allow the evaluation of the myoelectric activity on back muscles, the sEMG signals were filtered using a sixth-order Butterworth passband filter 30 Hz ÷ 450 Hz. The best value to filter ECG artifacts in sEMG signals [53] was 30 Hz, and 450 Hz was used to remove high-frequency harmonics [21,54]. The filtered sEMG signals (where signal exceed of the test end was discarded), the inclination, and the phases signal were superimposed providing the appropriate data to carry out the FRP analysis, as reported in the block diagram of Figure 3. Figure 5a–d shows the filtered sEMG signals, inclination signal, and phases signal processed by the proposed algorithm (they are referred to a control subject without LBP). Using this graphic representation, qualified medical personnel can easily carry out a visual inspection (VIS method) to analyse the back muscles activity and identify the presence/absence of FRP in each cycle of each sEMG channel. The greatest interest phase to make the decision, using the VIS method, is the full-flexion phase (number 3), where it's possible to observe FRP presence/absence.

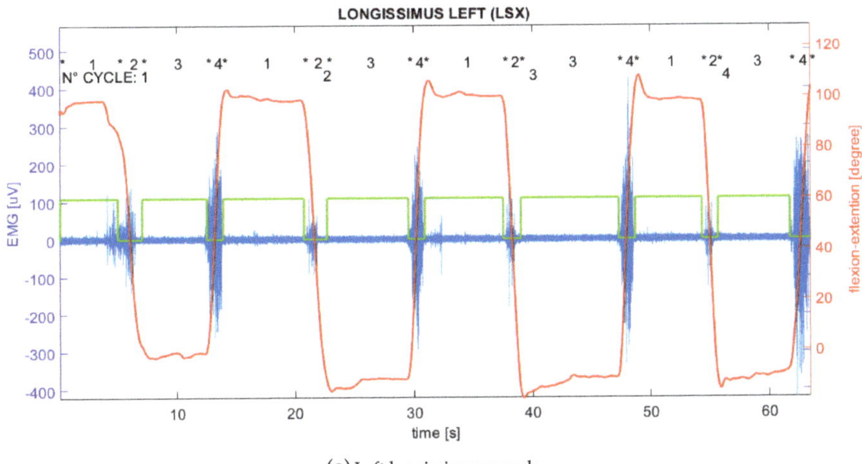

(**a**) Left longissimus muscle

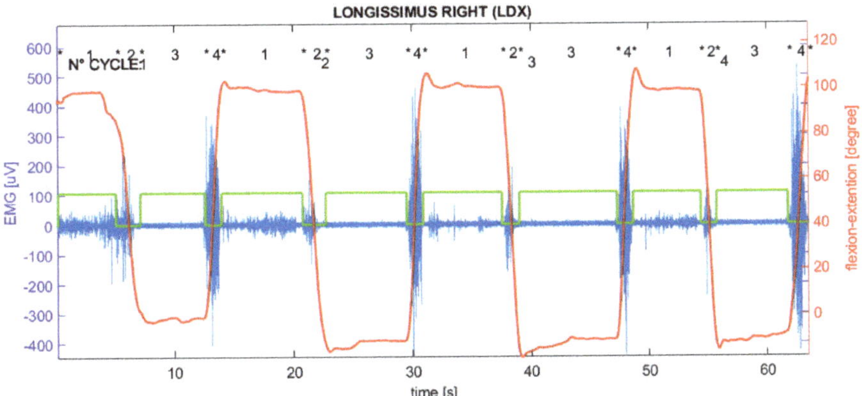

(**b**) Right longissimus muscle

Figure 5. *Cont.*

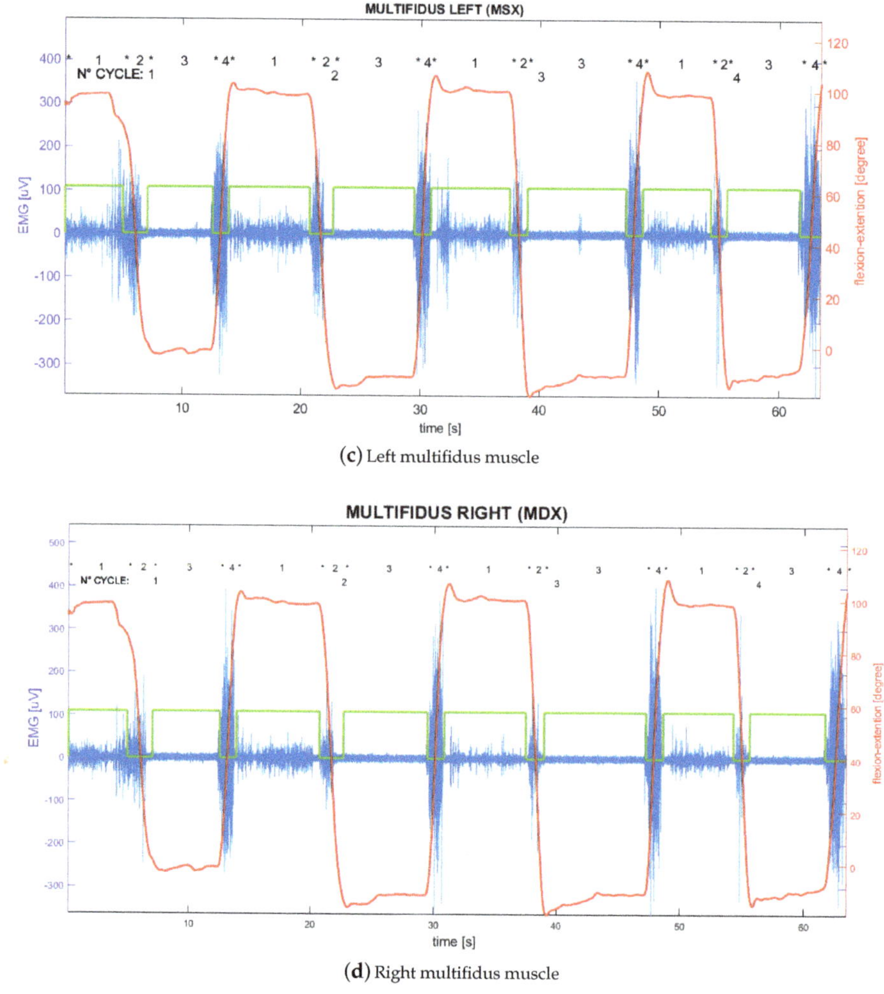

(**c**) Left multifidus muscle

(**d**) Right multifidus muscle

Figure 5. Graphic representation with the signals superimposition (filtered sEMG signal in blue, inclination signal in red, phases signal in green), phases (upper numbers) and cycles (lower numbers). It is referred to a healthy subject.

The second part of the algorithm consists to calculate the flexion-relaxation ratio (FRR) which is another technique, compared to the VIS method, to identify the FRP. The main difference is that the decision about FRP presence/absence in the VIS method is made by the doctor while using the FRR the final decision is made automatically by an algorithm. The goal was therefore to automate the decision-making process, to provide an objective contribution that could help the doctor in less time. Most of the FRRs used in the literature are the ratio of the EMG processed during the full-flexion phase on the numerator and EMG processed during the extension phase on the denominator. The reason is that theoretically under normal conditions, with healthy subjects, in the complete flexion we should obtain the lowest average activity thanks to FRP presence, while the average activity is higher, compared to all the other phases, during the extension (as is possible to see in Figure 6a–d); therefore, choosing a relationship between these two phases theoretically we have a greater gap (is easier to make the decision). For example, in the same conditions, if we choose the ratio between the full flexion and the standing phase the average gap is less (as is possible to see in Figure 6a–d).

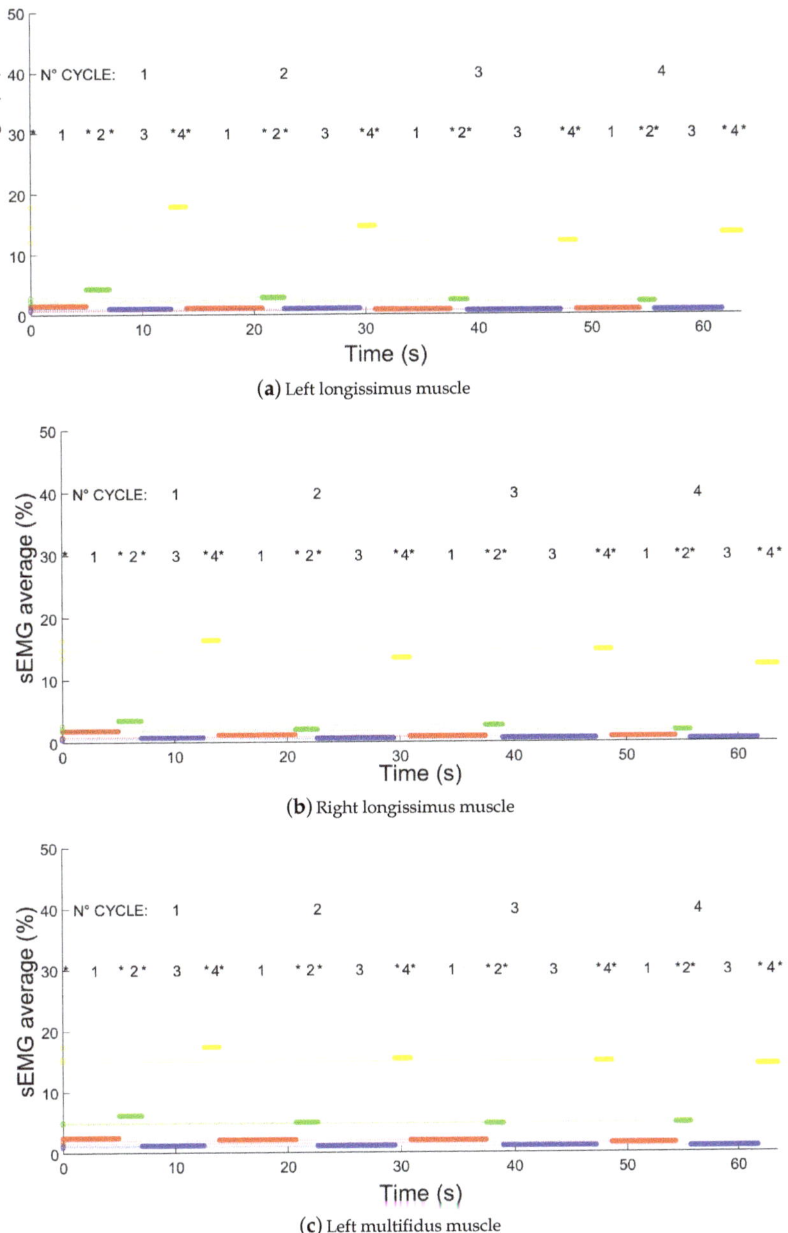

(**a**) Left longissimus muscle

(**b**) Right longissimus muscle

(**c**) Left multifidus muscle

Figure 6. *Cont.*

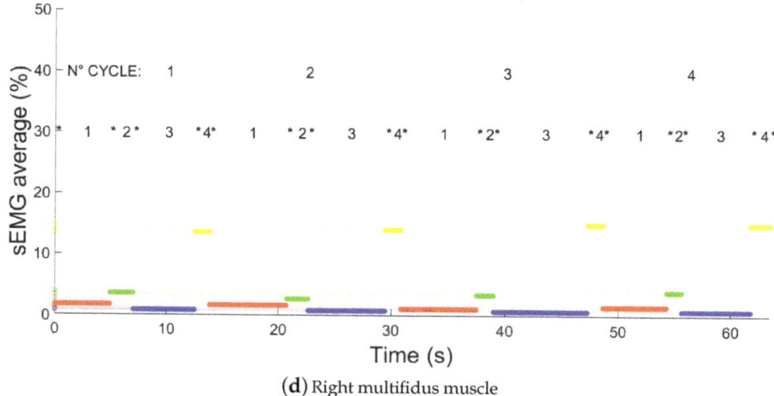

(**d**) Right multifidus muscle

Figure 6. sEMG filtered and rectified is normalized respect the max value of each cycle and the average sEMG levels, for each phase of each muscle, are expressed in percentage. Each phase is represented by a different colour: standing phase (red), flexion phase (green), full-flexion phase (blue), extension phase (yellow). They are referred to the same healthy subject of the previous graphs.

Calculate sEMG amplitudes ratios, between the motion phases, is a technique that allows normalization for repeated measures over time or for between-subject comparisons [12]. Ritvanenen et al. have computed the FRR by the ratio between the maximal RMS activity during 1 s of the flexion with the maximal RMS activity during 1 s of the full flexion; then the data were normalized by dividing them for the average sEMG activity during the standing phase [22]. Fernandes et al. have computed the FRR dividing the maximum RMS of EMG activity level during the flexion by the lowest mean EMG activity as measured over a 1-second interval during the full flexion phase [55]. In the literature there are many different types of FRRs, some of which have been explained and compared by Alison [20]. However, these studies did not report cutoff values to discriminate the presence/absence of FRP. Moreover, the comparison between FRR proposed in these studies is difficult due to the different factors used to evaluate the FRP and the lack of standardization. In order to overcome these limitations in this paper we propose an FRR method that uses a nominal threshold reference value to detect the presence or absence of FRP; this cutoff is an empirical value obtained during the analysis of the data collected in the acquisition campaign [43].

The specific FRR used in this study, also called flexion-extension ratio (FER), is the ratio between the average of the filtered and rectified sEMG signal during the full-flexion phase and the average of the filtered and rectified sEMG signal during the extension phase. This processing is made in the "FRR computation" block of Figure 3. The filtered sEMG signals were rectified and normalized respect to the max sEMG value in each cycle. Figure 7a–d shows the rectified and normalized sEMG signals where it is also illustrated the percentage angle (compared with the max angle) at which a new phase begins. These graphs are useful to evaluate the myoelectric activity on the back muscles and emphasizing the most common patterns: low activity in the standing phase, the activity increases in the flexion phase, myoelectric silence in full-flexion phase, and then the activity increases much more in the extension phase. As mentioned above, typically the myoelectric activity during the extension phase is greater than myoelectric activity during the flexion phase because it is necessary to contrast the torso strength, generated by the gravity acceleration, in opposition to the force generated by the muscles when they return to the starting position.

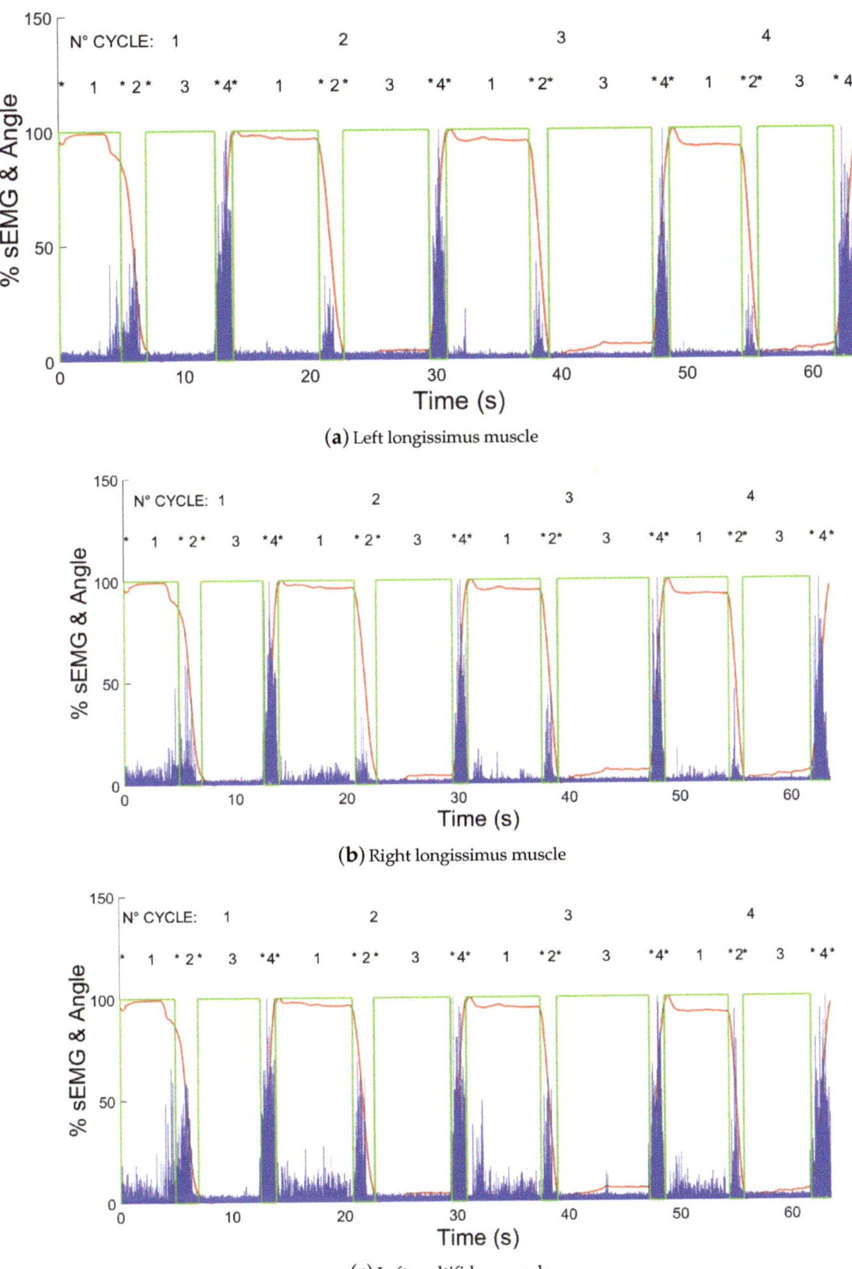

(**a**) Left longissimus muscle

(**b**) Right longissimus muscle

(**c**) Left multifidus muscle

Figure 7. *Cont.*

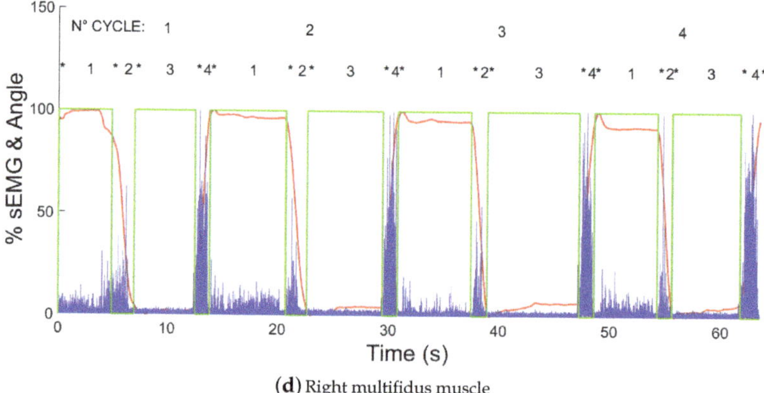

(d) Right multifidus muscle

Figure 7. Myoelectric activity for each phase and each muscle. Blue signal represents the sEMG normalized respect the max value of the cycle and it is expressed in percentage terms. The red signal is the inclination signal normalized respect the max value in the cycle and it is expressed in percentage terms. The green graph is the phases signal. They are referred to the same healthy subject of the previous graphs.

The processed sEMG signals were used to compute the FRR applying the following equation:

$$FRR_i^C = \frac{[\sum_{j=1}^{n}(|sEMG_j|/max(sEMG^i(t))) * 100]/n}{[\sum_{k=1}^{m}(|sEMG_k|/max(sEMG^i(t))) * 100]/m} \qquad For\ i = 1, \ldots, 4 \tag{1}$$

where:

- i = i-th cycle;
- j = j-th sample in full-flexion phase;
- k = k-th sample in extension phase;
- n = total samples in full-flexion phase;
- m = total samples in extension phase;
- $sEMG^i(t)$ = represents the total signal filtered and synchronized in the i-th cycle, where t is a discrete-time variable multiple of the sample time;
- $sEMG_j$ = represents the j-th amplitude of the sEMG signal (filtered and synchronized) in full flexion phase;
- $sEMG_k$ = represents the k-th amplitude of the sEMG signal (filtered and synchronized) in extension phase;
- C = type of Channel (LSX, LDX, MSX, MDX).

For each cycle, in Equation (1) the numerator is represented with the blue signal while the denominator is represented with the yellow signal, as reported in Figure 6a–d.

It should happen that when the FRP is present the full flexion activity is very low respect extension activity (ratio near zero), while when the FRP is absent the full flexion activity approaches the extension activity level (ratio near one). The last algorithm step consists to compare each $FRR_i{}^C$ with a nominal threshold reference value $FRR^{Threshold}$ set at 0.35. So, the final stage (detection block in Figure 3) considers the FRP presence if the calculated value is below 0.35 ($FRR_i{}^C < FRR^{Threshold}$) while it considers the FRP absence if the ratio is equal or greater than 0.35 ($FRR_i{}^C \geq FRR^{Threshold}$). The $FRR^{Threshold}$ was empirically estimated and we can define it as *"the best value that reduces better the differences between the VIS method and FRR method"*.

4. Results

The algorithm was tested using the dataset published in our previous study [43]. It contains acquisitions of 25 volunteer subjects; they repeated four times the cycle as defined in the flexion-relaxation test procedure [43]. In each cycle the electromyography signals of four muscles were acquired. Therefore, a total number of 400 events were extracted for the evaluation of the performances using the proposed algorithm for FRP clinical assessment. To identify if the automatic algorithm was taking the correct decision, in terms of FRP identification, a comparison between FRR and VIS method results were presented. VIS method was taken as a benchmark because of, due to its accurate performances, it was commonly adopted in clinical and research applications [20]. The recordings collected in the dataset were evaluated by independent blind teams composed of medical experts. Using VIS method (on the sEMG signals with the superimposed inclination signal), the blind teams provided handwritten reports in which the occurrences of FRP in each cycle have been reported. The results of the VIS method were obtained by summarizing the handwritten reports and in cases of disagreement between the three blind teams, a final decision was reached by the majority. The VIS method results are shown in Table 1, where each event of the dataset is classified by a Positive (P) or a Negative (N) outcome. In events with the positive outcome the blind teams have ascertained the FRP presence in the cycle under examination while in the events with a negative outcome the FRP absence was identified. The VIS method is based on criteria found in the literature: "A clear, sudden reduction in motor activity" [20].

Table 1. Results of the VIS method. For each subject, the cycles of each channel are classified as Positive (P) or Negative (N) events.

Subject ID	SEX	AGE	GROUP	LSX				LDX				MSX				MDX			
				1	2	3	4	1	2	3	4	1	2	3	4	1	2	3	4
1	F	51	LBP	P	P	P	P	P	P	P	P	P	P	P	P	P	P	P	P
2	F	40	HEALTHY	P	P	P	P	P	P	P	P	P	P	P	P	P	P	P	P
3	F	34	HEALTHY	P	P	P	P	P	P	P	P	P	P	P	P	P	P	P	P
4	M	57	LBP	N	P	P	P	N	P	P	P	N	N	N	N	N	N	P	P
5	M	30	LBP	N	N	N	N	N	N	N	N	N	N	N	N	N	N	N	N
6	M	31	HEALTHY	N	N	N	N	N	N	N	N	N	N	N	N	N	N	N	N
7	M	35	HEALTHY	P	P	P	P	P	P	P	P	N	P	P	P	P	P	P	P
8	M	25	HEALTHY	P	P	P	P	P	P	P	P	P	P	P	P	P	P	P	P
9	M	58	LBP	N	P	P	P	N	P	P	P	N	N	N	P	N	N	N	N
10	F	52	LBP	N	N	N	N	N	N	N	N	N	N	N	N	N	N	N	N
11	F	46	LBP	N	N	N	N	N	N	N	N	N	N	N	N	N	N	N	N
12	F	40	HEALTHY	P	P	P	P	P	P	P	P	P	P	P	P	P	P	P	P
13	M	49	LBP	N	N	N	N	P	P	P	P	N	N	N	N	N	N	N	N
14	F	49	LBP	P	P	P	P	P	P	P	P	N	N	N	N	N	N	N	N
15	F	51	LBP	N	P	P	P	N	P	P	P	N	N	N	N	N	N	N	N
16	F	60	HEALTHY	N	N	N	N	N	N	N	N	N	N	N	N	N	N	N	N
17	F	36	HEALTHY	P	P	P	P	P	P	P	P	P	P	P	P	P	P	P	P
18	M	22	HEALTHY	P	P	P	P	P	P	P	P	P	P	P	P	P	P	P	P
19	M	52	LBP	N	N	N	N	N	N	N	N	N	N	N	N	N	N	N	N
20	F	22	HEALTHY	P	P	P	P	P	P	P	P	P	P	P	P	P	P	P	P
21	M	60	HEALTHY	P	P	P	P	P	P	P	P	P	P	P	P	P	P	P	P
22	F	51	HEALTHY	N	N	N	N	N	N	N	N	N	N	N	N	N	N	N	N
23	M	60	LBP	N	N	N	N	N	N	N	N	N	N	N	N	N	N	N	N
24	M	61	LBP	N	N	N	N	N	N	N	N	N	N	N	N	N	N	P	P
25	M	52	HEALTHY	P	P	P	P	N	N	N	N	N	N	N	N	N	N	N	N

An ideal algorithm, for FRP detection, should identify the FRP onset in all the events with a positive outcome and it should recognize FRP absence in all the events with a negative outcome. Comparing the results of the VIS method with those obtained by the proposed automatic algorithm for FRP detection is possible to identify four types of events classification:

- True Positive (TP)—The algorithm correctly reported FRP presence in an event with a positive outcome;
- False Positive (FP)—The algorithm incorrectly reported FRP presence in an event with a negative outcome;
- True Negative (TN)—The algorithm correctly reported FRP absence in an event with a negative outcome;
- False Negative (FN)—The algorithm incorrectly reported FRP absence in an event with a positive outcome.

The information, collected in the dataset, was processed by the proposed algorithm for FRP detection and the results have been shown together with the flexion-extension ratio values calculated in each cycle and channel (Table 2), using Equation (1). Table 3 shows the mean and the SD obtained using the algorithm for computing ratios between full-flexion and extension phase.

Since the number of dataset events was statistically significant, it was possible to derive accuracy, sensitivity, and specificity about the performances of the proposed algorithm using the following equations:

$$A_c = \frac{TP+TN}{P+N} = \frac{TP+TN}{TP+FP+TN+FN} = \frac{382}{400} = 95.5\% \tag{2}$$

$$S_e = \frac{TP}{P} = \frac{TP}{TP+FN} = \frac{195}{195+3} = 98.5\% \tag{3}$$

$$S_p = \frac{TN}{N} = \frac{TN}{TN+FP} = \frac{187}{187+15} = 92.6\% \tag{4}$$

Table 2. Results of the FRR method for all 400 events. Each value contains the event classification and the relative ratio. It was used an $FRR^{Threshold} = 0.35$.

Subject ID	LSX				LDX				MSX				MDX			
	1	2	3	4	1	2	3	4	1	2	3	4	1	2	3	4
1	TP-0.07	TP-0.07	TP-0.07	TP-0.07	TP-0.08	TP-0.08	TP-0.08	TP-0.08	TP-0.09	TP-0.07	TP-0.08	TP-0.08	TP-0.08	TP-0.08	TP-0.08	TP-0.08
2	TP-0.08	TP-0.08	TP-0.08	TP-0.09	TP-0.08	TP-0.08	TP-0.11	TP-0.06	TP-0.08	TP-0.07	TP-0.08	TP-0.07	TP-0.09	TP-0.09	TP-0.09	TP-0.08
3	TP-0.06	TP-0.07	TP-0.06	TP-0.05	TP-0.05	TP-0.05	TP-0.05	TP-0.04	TP-0.07	TP-0.08	TP-0.07	TP-0.07	TP-0.06	TP-0.06	TP-0.05	TP-0.06
4	TN-0.47	TP-0.20	TP-0.15	TP-0.18	TN-0.36	TP-0.14	TP-0.10	TP-0.11	TN-0.67	TN-0.37	FP-0.26	FP-0.20	TN-0.83	FP-0.32	TP-0.18	TP-0.17
5	TN-0.71	TN-0.58	TN-0.50	TN-0.35	TN-0.79	TN-0.65	TN-0.57	TN-0.37	TN-0.67	TN-0.61	TN-0.55	TN-0.36	TN-0.81	TN-0.74	TN-0.68	TN-0.41
6	TN-0.61	TN-0.58	TN-0.89	TN-0.63	TN-0.59	TN-0.51	TN-0.84	TN-0.69	TN-0.43	TN-0.36	TN-0.67	TN-0.49	TN-0.35	FP-0.33	TN-0.48	TN-0.35
7	TP-0.19	TP-0.15	TP-0.13	TP-0.13	TP-0.21	TP-0.19	TP-0.16	TP-0.20	TN-0.38	TP-0.28	TP-0.24	TP-0.21	TP-0.33	TP-0.28	TP-0.26	TP-0.28
8	TP-0.08	TP-0.08	TP-0.10	TP-0.08	TP-0.17	TP-0.14	TP-0.18	TP-0.16	TP-0.10	TP-0.10	TP-0.12	TP-0.10	TP-0.14	TP-0.13	TP-0.15	TP-0.15
9	FP-0.18	TP-0.13	TP-0.11	TP-0.14	FP-0.25	TP-0.16	TP-0.13	TP-0.17	TN-0.58	TN-0.40	TN-0.36	TP-0.30	TN-0.54	TN-0.42	TN-0.42	TN-0.35
10	TN-0.74	TN-0.84	TN-0.85	TN-1.27	TN-0.65	TN-0.91	TN-0.92	TN-1.25	TN-0.88	TN-0.90	TN-0.79	TN-1.16	TN-0.73	TN-0.91	TN-0.78	TN-1.11
11	TN-0.97	TN-1.34	TN-0.76	TN-0.42	TN-1.07	TN-1.11	TN-1.00	TN-0.52	TN-1.08	TN-1.58	TN-1.00	TN-0.50	TN-1.06	TN-1.21	TN-0.99	TN-0.49
12	TP-0.10	TP-0.07	TP-0.06	TP-0.06	TP-0.11	TP-0.09	TP-0.10	TP-0.09	TP-0.07	TP-0.04	TP-0.04	TP-0.04	TP-0.09	TP-0.07	TP-0.08	TP-0.07
13	TN-0.57	TN-0.41	TN-0.36	TN-0.39	TP-0.28	TP-0.25	TP-0.17	TP-0.23	TN-0.68	TN-0.62	TN-0.70	TN-0.62	TN-0.61	TN-0.56	TN-0.63	TN-0.55
14	TP-0.24	TP-0.15	TP-0.25	FN-1.04	TP-0.25	TP-0.18	TP-0.32	FN-0.90	TN-0.60	TN-0.40	TN-0.52	TN-0.97	TN-0.71	TN-0.52	TN-0.55	TN-0.83
15	TN-0.45	TP-0.26	TP-0.30	TP-0.22	TN-0.49	TP-0.34	TP-0.33	TP-0.29	TN-0.68	TN-0.48	TN-0.52	TN-0.40	TN-0.68	TN-0.52	TN-0.52	TN-0.44
16	TN-0.48	TN-0.44	FP-0.33	FP-0.32	TN-0.52	TN-0.43	TN-0.37	TN-0.39	TN-1.00	TN-0.92	TN-0.84	TN-0.86	TN-0.80	TN-0.64	TN-0.54	TN-0.47
17	TP-0.13	TP-0.11	TP-0.14	TP-0.15	TP-0.22	TP-0.20	TP-0.18	TP-0.24	TP-0.19	TP-0.17	TP-0.21	TP-0.23	TP-0.26	TP-0.22	TP-0.24	TP-0.28
18	TP-0.08	TP-0.06	TP-0.06	TP-0.06	TP-0.13	TP-0.09	TP-0.09	TP-0.10	TP-0.06	TP-0.04	TP-0.04	TP-0.03	TP-0.18	TP-0.05	TP-0.05	TP-0.05
19	TN-0.83	TN-0.69	TN-0.60	TN-0.64	TN-0.65	TN-0.50	TN-0.43	TN-0.39	TN-0.75	TN-0.71	TN-0.62	TN-0.79	TN-0.73	TN-0.67	TN-0.62	TN-0.72
20	TP-0.13	TP-0.14	TP-0.14	TP-0.22	TP-0.12	TP-0.11	TP-0.15	TP-0.23	TP-0.09	TP-0.08	TP-0.10	TP-0.14	TP-0.07	TP-0.07	TP-0.09	TP-0.12
21	TP-0.32	TP-0.25	TP-0.26	TP-0.22	TP-0.15	TP-0.13	TP-0.15	TP-0.13	TP-0.20	TP-0.19	TP-0.21	TP-0.21	TP-0.21	TP-0.22	TP-0.29	TP-0.21
22	TN-0.57	TN-0.58	TN-0.53	FP-0.33	TN-0.58	TN-0.58	TN-0.58	FP-0.34	TN-0.70	TN-0.48	TN-0.49	TN-0.36	TN-0.76	TN-0.43	TN-0.37	FP-0.32
23	TN-0.45	TN-0.56	TN-0.56	TN-0.36	TN-0.86	TN-0.85	TN-0.80	TN-0.68	TN-1.16	TN-1.36	TN-1.18	TN-1.30	TN-0.74	TN-0.79	TN-0.72	TN-0.75
24	TN-0.67	TN-0.61	TN-0.53	TN-0.87	TN-0.61	TN-0.59	TN-0.46	TN-0.66	TN-0.62	TN-0.60	TN-0.38	TN-0.61	TN-0.46	TN-0.49	TP-0.20	FN-0.55
25	TP-0.28	TP-0.21	TP-0.21	TP-0.25	TN-0.49	TN-0.53	TN-0.58	TN-0.55	TN-0.55	TN-0.62	TN-0.66	TN-0.64	FP-0.34	FP-0.26	FP-0.27	FP-0.28

Table 3. Mean and standard deviation of the FRR, for Healthy and LBP subjects, obtained using the proposed method.

GROUP	FRR
HEALTHY	0.25 ± 0.20
LBP	0.55 ± 0.26

5. Discussion

The results of the VIS method (Table 1) and FRR method (Table 2) have shown that a subset of subjects exhibited FRP in all cycles and in all muscles, and it was indicated with ID^{FRP}= *(ID1, ID2, ID3, ID8, ID12, ID17, ID18, ID20, ID21)*. The other subjects manifested FRP only in some cycles, some muscles or they didn't manifest FRP in any muscles. The reason for different patterns may lie in muscle fatigue [56], fear, or other features that vary from subject to subject (since there are many variables involved; the patient's report is useful for deepening the topic). Comparing the results of the VIS method with those obtained by the proposed algorithm is possible to carry out the performance evaluation, which is summarized in Table 2. Taking into account only the results related to the subgroup of healthy subjects indicated with $ID^{HEALTHY}$= *(ID2, ID3, ID6, ID7, ID8, ID12, ID16, ID17, ID18, ID20, ID21, ID22, ID25)*. The proposed FRR algorithm shows no False Negative, so:

$$S_e^{HEALTHY} = \frac{TP}{P} = \frac{TP}{TP+FN} = \frac{147}{147+0} = 100\% \tag{5}$$

$$S_p^{HEALTHY} = \frac{TN}{N} = \frac{TN}{TN+FP} = \frac{51}{51+10} = 84\% \tag{6}$$

This means that the ability of the FRR method, Equation (1), to discriminate FRP presence, compared to the VIS method, in the healthy subgroup (with a total number of 147 events) is equal to 100% (Equation (5)). In the literature, Alison et al. [20] reported a sensibility of 100% evaluating other types of FRR methods on 24 events collected in an acquisition campaign that involved only healthy subjects. Therefore, we have obtained similar performance comparing our results with those reported by Alison in the same type of subset but using a greater number of events. While as reported in Equation (6), the proposed FRR algorithm is less performing to discriminate FRP absence in the healthy subgroup (with a total number of 61 events).

We can do similar considerations taking into account only the results related to the subgroup of LBP subjects indicated with ID^{LBP} = *(ID1, ID4, ID5, ID9, ID10, ID11, ID13, ID14, ID15, ID19, ID23, ID24)*. The proposed algorithm shows:

$$S_e^{LBP} = \frac{TP}{P} = \frac{TP}{TP+FN} = \frac{48}{48+3} = 94.1\% \tag{7}$$

$$S_p^{LBP} = \frac{TN}{N} = \frac{TN}{TN+FP} = \frac{136}{136+5} = 96.5\% \tag{8}$$

This means that the ability of the FRR method, Equation (1), to discriminate FRP presence, compared to the VIS method, in LBP subgroup (with a total number of 51 events) is equal to 94.1% (Equation (7)). While as reported in Equation (8), the proposed FRR algorithm is more performing to discriminate FRP absence in the LBP subgroup (with a total number of 141 events).

Summarizing, in this study we have extended the evaluation of the FRP taking subjects healthy and with LBP, producing 400 events, and evaluating the performances of the proposed algorithm on the entire group of subjects. The results are illustrated in Table 2, and the performances are indicated in Equations (2)–(4). The table shows that this algorithm correctly recognized 195 of the 198 events with FRP and 187 of the 202 events without FRP. Only 15 False Positives and three False Negatives occurred on a total of 400 events. FNs can occur in events where the end of a phase does not exactly coincide with the end of the sEMG pattern for that phase. Then the sEMG pattern excess enters in the new phase and it alters the average value, changing the FRR. FPs can occur in events where the FRP is absent and the sEMG activity is moderate and very variable during the full flexion phase. In these cases, the FRR value does not exceed the nominal threshold as the mean value during the full flexion phase is not big enough compared to the activity during the extension.

As shown in equations Equations (2)–(4), the algorithm has therefore accuracy of 95.5%, a sensitivity of 98.5%, and a specificity of 92.6%. Comparing sensitivity Equation (3) and specificity Equation (4) results is clear that the algorithm discriminates very well subjects with FRP while it's more difficult to identify subjects without FRP. Furthermore, as reported in Equation (6), this reduced specificity is mainly caused by healthy subjects who have not FRP.

Most cases of FPs had a value very near to the threshold level ($FRR^{Threshold}$) used to make the decision; often, when the FRR_i^C value is near the threshold it can take a different decision compared to the VIS method (the closer FRR gets to the threshold, the greater the uncertainty of the decision), which makes the detection a non-trivial problem.

6. Conclusions

This paper deepens the investigation of FRP on back muscles using a WBSN composed of four sEMG sensors and a wearable device that integrates accelerometer, gyroscope, and magnetometer. The raw data collected from the WBSN during the flexion-relaxation test are processed by an algorithm able to identify the phases of which the test is composed, provide an evaluation of the myoelectric activity and automatically detect the FRP presence/absence. The proposed algorithm was tested using the data acquired in an acquisition campaign conducted to evaluate the flexion-relaxation phenomenon on the back muscles of subjects with and without LBP. The computed signal, identifying the phases of the flexion relaxation test, represented very well the subject's trunk real motion. Moreover, the phases signal trend varies in correspondence of the angular variations causing the activation or not of the muscles. The assessment of the myoelectric activity on back muscles provided by the proposed algorithm was evaluated by the medical staff as a useful tool to identify and cluster different patterns, visually analyse FRP presence/absence with the VIS method and aid the clinical assessment of the FRP. The ratio, expressed by Equation (1), was computed for each event using the data collected in the acquisition campaign. This FRR parameter is then compared with an empirical threshold value and the final decision about flexion-relaxation phenomenon presence/absence is taken. The threshold level used in the algorithm to detect FRP seemed to classify very well the events collected in the dataset. Indeed, the results show that the proposed algorithm for the FRP detection obtained an accuracy of 95.5%, a sensitivity of 98.5%, and a specificity of 92.6%, processing the data acquired from the subjects with and without LBP. Despite the excellent results achieved, future developments will concern the planning of a new acquisition campaign and the study of new solutions able to detect the FRP and to improve the performance of the proposed algorithm. In the future acquisition campaign, different motion tasks will be taken into consideration to evaluate the FRP and the psycho-physical conditions of the subjects involved (health conditions, level of stress, etc.) will be carefully monitored before carrying them out. Moreover, the relationship between the FRR method and the low back pain will be studied in future works in order to try to discriminate healthy subjects from LBP patients by analysing FRRs parameters without knowing a priori the clinical conditions.

Author Contributions: Conceptualization, M.P., A.B., L.P. and P.P.; methodology, M.P. and P.P.; software, M.P. and A.B.; validation, M.P. and L.P.; formal analysis, M.P. and M.V.; investigation, M.P.; resources M.P., P.P. and M.V.; data curation, M.P.; writing–original draft preparation, M.P. and A.B.; writing–review and editing, M.P. and A.B.; visualization, M.P and A.B; supervision, P.P.; project administration, P.P.; funding acquisition, P.P. All authors have read and agreed to the published version of the manuscript.

Funding: This research was funded by Department of Information Engineering (DII), Università Politecnica delle Marche, project title "A network-based approach to uniformly extract knowledge and support decision making in heterogeneous application contexts", grant number RSA-B2018.

Acknowledgments: The authors are grateful to the Istituto di Riabilitazione Santo Stefano (www.sstefano.it) for the opportunity to carry out this research work and the technical support of the materials used for experiments.

Conflicts of Interest: The authors declare no conflict of interest.

Abbreviations

The following abbreviations are used in this manuscript:

ACC	Acceleration
BSN	Body Sensor Network
ECG	Electrocardiogram
FN	False Negative
FP	False Positive
FER	Flexion Extension Ratio
FRP	Flexion Relaxation Phenomenon
FRR (or plural FRRs)	Flexion Relaxation Ratio (s)
GYR	Gyroscope
ID	Identification Number
LBP	Low Back Pain
LDX	Longissimus Right
LSX	Longissimus Left
MAG	Magnetic field
MDX	Multifidus Right
MSX	Multifidus Left
NRS-11	Numeric Rating Scale
RMS	Root Mean Square
sEMG	Surface Electromyography
TP	True Positive
TN	True Negative
VIS	Visual Inspection
WBSN	Wireless Body Sensor Network

References

1. Floyd, W.; Silver, P. The function of the erectores spinae muscles in certain movements and postures in man. *J. Physiol.* **1955**, *129*, 184–203. [CrossRef] [PubMed]
2. McGill, S.M.; Kippers, V. Transfer of loads between lumbar tissues during the flexion-relaxation phenomenon. *Spine* **1994**, *19*, 2190–2196. [CrossRef] [PubMed]
3. Andersson, E.; Oddsson, L.; Grundström, H.; Nilsson, J.; Thorstensson, A. EMG activities of the quadratus lumborum and erector spinae muscles during flexion-relaxation and other motor tasks. *Clin. Biomech.* **1996**, *11*, 392–400. [CrossRef]
4. Hodges, P.W.; Moseley, G.L. Pain and motor control of the lumbopelvic region: Effect and possible mechanisms. *J. Electromyogr. Kinesiol.* **2003**, *13*, 361–370. [CrossRef]
5. Hodges, P.W.; Tucker, K. Moving differently in pain: A new theory to explain the adaptation to pain. *Pain* **2011**, *152*, S90–S98. [CrossRef]
6. van Dieën, J.H.; Selen, L.P.; Cholewicki, J. Trunk muscle activation in low-back pain patients, an analysis of the literature. *J. Electromyogr. Kinesiol.* **2003**, *13*, 333–351. [CrossRef]
7. Golding, J. Electromyography of the erector spinae in low back pain. *Postgrad. Med. J.* **1952**, *28*, 401. [CrossRef]
8. Bhms, V.K.; Parker, A. Posture related to myoelectric silence of erectores spinae during trunk flexion. *Spine* **1984**, *9*, 740.
9. Sarti, M.; Lison, J.; Monfort, M.; Fuster, M. Response of the flexion–relaxation phenomenon relative to the lumbar motion to load and speed. *Spine* **2001**, *26*, E421–E426. [CrossRef] [PubMed]

10. Watson, P.J.; Booker, C.; Main, C.; Chen, A. Surface electromyography in the identification of chronic low back pain patients: The development of the flexion relaxation ratio. *Clin. Biomech.* **1997**, *12*, 165–171. [CrossRef]
11. Sihvonen, T.; Partanen, J.; Hänninen, O.; Soimakallio, S. Electric behavior of low back muscles during lumbar pelvic rhythm in low back pain patients and healthy controls. *Arch. Phys. Med. Rehabil.* **1991**, *72*, 1080–1087. [PubMed]
12. McGorry, R.W.; Lin, J.H. Flexion relaxation and its relation to pain and function over the duration of a back pain episode. *PLoS ONE* **2012**, *7*, e39207. [CrossRef] [PubMed]
13. Neblett, R. Surface electromyographic (SEMG) biofeedback for chronic low back pain. *Healthcare* **2016**, *4*, 27. [CrossRef]
14. Othman, S.H.; Ibrahim, F.; Omar, S.; Rahim, R. Flexion relaxation phenomenon of back muscles in discriminating between healthy and chronic low back pain women. In Proceedings of the 4th Kuala Lumpur International Conference on Biomedical Engineering 2008, Kuala Lumpur, Malaysia, 25–28 June 2008; pp. 199–203.
15. Alschuler, K.N.; Neblett, R.; Wiggert, E.; Haig, A.J.; Geisser, M.E. Flexion-relaxation and clinical features associated with chronic low back pain: A comparison of different methods of quantifying flexion-relaxation. *Clin. J. Pain* **2009**, *25*, 760–766. [CrossRef] [PubMed]
16. Geisser, M.E.; Ranavaya, M.; Haig, A.J.; Roth, R.S.; Zucker, R.; Ambroz, C.; Caruso, M. A meta-analytic review of surface electromyography among persons with low back pain and normal, healthy controls. *J. Pain* **2005**, *6*, 711–726. [CrossRef] [PubMed]
17. Neblett, R.; Mayer, T.G.; Gatchel, R.J.; Proctor, T. Quantifying lumbar flexion-relaxation phenomenon: Theory and clinical applications. *Spine J.* **2002**, *2*, 97. [CrossRef]
18. Schinkel-Ivy, A.; Nairn, B.C.; Drake, J.D. Quantification of the lumbar flexion-relaxation phenomenon: Comparing outcomes of lumbar erector spinae and superficial lumbar multifidus in standing full trunk flexion and slumped sitting postures. *J. Manip. Physiol. Ther.* **2014**, *37*, 494–501. [CrossRef]
19. Jin, S.; Ning, X.; Mirka, G.A. An algorithm for defining the onset and cessation of the flexion-relaxation phenomenon in the low back musculature. *J. Electromyogr. Kinesiol.* **2012**, *22*, 376–382. [CrossRef]
20. Schinkel-Ivy, A.; Nairn, B.C.; Drake, J.D. Evaluation of methods for the quantification of the flexion-relaxation phenomenon in the lumbar erector spinae muscles. *J. Manip. Physiol. Ther.* **2013**, *36*, 349–358. [CrossRef]
21. Nougarou, F.; Massicotte, D.; Descarreaux, M. Detection method of flexion relaxation phenomenon based on wavelets for patients with low back pain. *EURASIP J. Adv. Signal Process.* **2012**, *2012*, 151. [CrossRef]
22. Ritvanen, T.; Zaproudina, N.; Nissen, M.; Leinonen, V.; Hänninen, O. Dynamic surface electromyographic responses in chronic low back pain treated by traditional bone setting and conventional physical therapy. *J. Manip. Physiol. Ther.* **2007**, *30*, 31–37. [CrossRef] [PubMed]
23. Sihvonen, T. Flexion relaxation of the hamstring muscles during lumbar-pelvic rhythm. *Arch. Phys. Med. Rehabil.* **1997**, *78*, 486–490. [CrossRef]
24. CARLSÖÖ, S. The static muscle load in different work positions: An electromyographic study. *Ergonomics* **1961**, *4*, 193–211. [CrossRef]
25. Mayer, T.; Tencer, A.F.; Kristoferson, S.; Mooney, V. Use of noninvasive techniques for quantification of spinal range-of-motion in normal subjects and chronic low-back dysfunction patients. *Spine* **1984**, *9*, 588–595. [CrossRef]
26. Mayer, T.G.; Neblett, R.; Brede, E.; Gatchel, R.J. The quantified lumbar flexion-relaxation phenomenon is a useful measurement of improvement in a functional restoration program. *Spine* **2009**, *34*, 2458–2465. [CrossRef]
27. Leardini, A.; Biagi, F.; Belvedere, C.; Benedetti, M.G. Quantitative comparison of current models for trunk motion in human movement analysis. *Clin. Biomech.* **2009**, *24*, 542–550. [CrossRef]
28. Solomonow, M.; Baratta, R.V.; Banks, A.; Freudenberger, C.; Zhou, B.H. Flexion–relaxation response to static lumbar flexion in males and females. *Clin. Biomech.* **2003**, *18*, 273–279. [CrossRef]
29. Chen, Y.L.; Chen, Y.; Lin, W.C.; Liao, Y.H.; Lin, C.J. Lumbar posture and individual flexibility influence back muscle flexion-relaxation phenomenon while sitting. *Int. J. Ind. Ergon.* **2019**, *74*, 102840. [CrossRef]
30. Roghani, T.; Zavieh, M.K.; Talebian, S.; Baghban, A.A.; Katzman, W. Back Muscle Function in Older Women With Age-Related Hyperkyphosis: A Comparative Study. *J. Manip. Physiol. Ther.* **2019**, *42*, 284–294. [CrossRef]

31. Goodvin, C.; Park, E.J.; Huang, K.; Sakaki, K. Development of a real-time three-dimensional spinal motion measurement system for clinical practice. *Med Biol. Eng. Comput.* **2006**, *44*, 1061–1075. [CrossRef]
32. Wong, W.Y.; Wong, M.S. Trunk posture monitoring with inertial sensors. *Eur. Spine J.* **2008**, *17*, 743–753. [CrossRef] [PubMed]
33. Ronchi, A.J.; Lech, M.; Taylor, N.; Cosic, I. A reliability study of the new Back Strain Monitor based on clinical trials. In Proceedings of the 2008 30th Annual International Conference of the IEEE Engineering in Medicine and Biology Society, Vancouver, BC, Canada, 20–24 August 2008; pp. 693–696.
34. Valencia-Jimenez, N.; Leal-Junior, A.; Avellar, L.; Vargas-Valencia, L.; Caicedo-Rodríguez, P.; Ramírez-Duque, A.A.; Lyra, M.; Marques, C.; Bastos, T.; Frizera, A. A Comparative Study of Markerless Systems Based on Color-Depth Cameras, Polymer Optical Fiber Curvature Sensors, and Inertial Measurement Units: Towards Increasing the Accuracy in Joint Angle Estimation. *Electronics* **2019**, *8*, 173. [CrossRef]
35. Mjøsund, H.L.; Boyle, E.; Kjaer, P.; Mieritz, R.M.; Skallgård, T.; Kent, P. Clinically acceptable agreement between the ViMove wireless motion sensor system and the Vicon motion capture system when measuring lumbar region inclination motion in the sagittal and coronal planes. *BMC Musculoskelet. Disord.* **2017**, *18*, 124. [CrossRef] [PubMed]
36. Kumar, P.; Saini, R.; Tumma, C.S.; Roy, P.P.; Dogra, D.P. Gait Analysis Using Shadow Motion. In Proceedings of the 2017 4th IAPR Asian Conference on Pattern Recognition (ACPR), Nanjing, China, 26–29 November 2017; pp. 453–458.
37. Kumar, P.; Mukherjee, S.; Saini, R.; Kaushik, P.; Roy, P.P.; Dogra, D.P. Multimodal gait recognition with inertial sensor data and video using evolutionary algorithm. *IEEE Trans. Fuzzy Syst.* **2018**, *27*, 956–965. [CrossRef]
38. Bergmann, J.H.; Chandaria, V.; McGregor, A. Wearable and implantable sensors: The patient's perspective. *Sensors* **2012**, *12*, 16695–16709. [CrossRef]
39. Mokhlespour Esfahani, M.I.; Nussbaum, M.A. Preferred placement and usability of a smart textile system vs. inertial measurement units for activity monitoring. *Sensors* **2018**, *18*, 2501. [CrossRef]
40. Rose-Dulcina, K.; Genevay, S.; Dominguez, D.; Armand, S.; Vuillerme, N. Flexion-Relaxation Ratio Asymmetry and Its Relation With Trunk Lateral ROM in Individuals With and Without Chronic Nonspecific Low Back Pain. *Spine* **2020**, *45*, E1–E9. [CrossRef]
41. Carrillo-Perez, F.; Diaz-Reyes, I.; Damas, M.; Banos, O.; Soto-Hermoso, V.M.; Molina-Molina, A. A Novel Automated Algorithm for Computing Lumbar Flexion Test Ratios Enhancing Athletes Objective Assessment of Low Back Pain. In Proceedings of the 2018 6th International Congress on Sport Sciences Research and Technology Support, Seville, Spain, 20–21 September 2018; pp. 34–39.
42. Laird, R.A.; Keating, J.L.; Ussing, K.; Li, P.; Kent, P. Does movement matter in people with back pain? Investigating 'atypical' lumbo-pelvic kinematics in people with and without back pain using wireless movement sensors. *BMC Musculoskelet. Disord.* **2019**, *20*, 28. [CrossRef]
43. Paoletti, M.; Belli, A.; Palma, L.; Paniccia, M.; Tombolini, F.; Ruggiero, A.; Vallasciani, M.; Pierleoni, P. Data acquired by wearable sensors for the evaluation of the flexion-relaxation phenomenon. *Data Brief* **2020**, under review.
44. Morbidoni, C.; Cucchiarelli, A.; Fioretti, S.; Di Nardo, F. A deep learning approach to EMG-based classification of gait phases during level ground walking. *Electronics* **2019**, *8*, 894. [CrossRef]
45. Hermens, H.J.; Freriks, B.; Merletti, R.; Stegeman, D.; Blok, J.; Rau, G.; Disselhorst-Klug, C.; Hägg, G. European recommendations for surface electromyography. *Roessingh Res. Dev.* **1999**, *8*, 13–54.
46. Li, Z.; Guan, X.; Zou, K.; Xu, C. Estimation of Knee Movement from Surface EMG Using Random Forest with Principal Component Analysis. *Electronics* **2020**, *9*, 43. [CrossRef]
47. Acar, G.; Ozturk, O.; Golparvar, A.J., Elboshra, T.A.; Böhringer, K.; Yapici, M.K. Wearable and flexible textile electrodes for biopotential signal monitoring: A review. *Electronics* **2019**, *8*, 479. [CrossRef]
48. Hartrick, C.T.; Kovan, J.P.; Shapiro, S. The numeric rating scale for clinical pain measurement: A ratio measure? *Pain Pract.* **2003**, *3*, 310–316. [CrossRef] [PubMed]
49. Tesio, L.; Granger, C.V.; Fiedler, R.C. A unidimensional pain/disability measure for low-back pain syndromes. *Pain* **1997**, *69*, 269–278. [CrossRef]
50. Madgwick, S.O.; Harrison, A.J.; Vaidyanathan, R. Estimation of IMU and MARG orientation using a gradient descent algorithm. In Proceedings of the 2011 IEEE International Conference on Rehabilitation Robotics, Zurich, Switzerland, 29 June–1 July 2011; pp. 1–7.

51. Pierleoni, P.; Belli, A.; Palma, L.; Pernini, L.; Valenti, S. An accurate device for real-time altitude estimation using data fusion algorithms. In Proceedings of the 2014 IEEE/ASME 10th International Conference on Mechatronic and Embedded Systems and Applications (MESA), Ancona, Italy, 10–12 September 2014; pp. 1–5.
52. Sabatini, A.M. Estimating three-dimensional orientation of human body parts by inertial/magnetic sensing. *Sensors* **2011**, *11*, 1489–1525. [CrossRef] [PubMed]
53. Redfern, M.S.; Hughes, R.E.; Chaffin, D.B. High-pass filtering to remove electrocardiographic interference from torso EMG recordings. *Clin. Biomech.* **1993**, *8*, 44–48. [CrossRef]
54. Sánchez-Zuriaga, D.; López-Pascual, J.; Garrido-Jaén, D.; García-Mas, M.A. A comparison of lumbopelvic motion patterns and erector spinae behavior between asymptomatic subjects and patients with recurrent low back pain during pain-free periods. *J. Manip. Physiol. Ther.* **2015**, *38*, 130–137. [CrossRef]
55. Fernandes, W.V.B.; Blanco, C.R.; Politti, F.; de Cordoba Lanza, F.; Lucareli, P.R.G.; Corrêa, J.C.F. The effect of a six-week osteopathic visceral manipulation in patients with non-specific chronic low back pain and functional constipation: study protocol for a randomized controlled trial. *Trials* **2018**, *19*, 151. [CrossRef]
56. Descarreaux, M.; Lafond, D.; Jeffrey-Gauthier, R.; Centomo, H.; Cantin, V. Changes in the flexion relaxation response induced by lumbar muscle fatigue. *BMC Musculoskelet. Disord.* **2008**, *9*, 10. [CrossRef]

Article

The Influence of Different Levels of Cognitive Engagement on the Seated Postural Sway

Daniele Bibbo *, Silvia Conforto, Maurizio Schmid and Federica Battisti

Department of Engineering, University of Roma Tre, Via Vito Volterra, 62, 00146 Rome, Italy; silvia.conforto@uniroma3.it (S.C.); maurizio.schmid@uniroma3.it (M.S.); federica.battisti@uniroma3.it (F.B.)

* Correspondence: daniele.bibbo@uniroma3.it; Tel.: +39-06-5733-7298

Received: 5 March 2020; Accepted: 30 March 2020; Published: 31 March 2020

Abstract: In this paper, we introduced and tested a new system based on a sensorized seat, to evaluate the sitting dynamics and sway alterations caused by different cognitive engagement conditions. An office chair was equipped with load cells, and a digital and software interface was developed to extract the Center of Pressure (COP). A population of volunteers was recruited to evaluate alterations to their seated posture when undergoing a test specifically designed to increase the cognitive engagement and the level of stress. Relevant parameters of postural sway were extracted from the COP data, and significant alterations were found in all of them, highlighting the ability of the system to capture the emergence of a different dynamic behavior in postural control when increasing the complexity of the cognitive engagement. The presented system can thus be used as a valid and reliable instrument to monitor the postural patterns of subjects involved in tasks performed in a seated posture, and this may prove useful for a variety of applications, including those associated with improving the quality of working conditions.

Keywords: seated posture; cognitive engagement; stress level; load cells; embedded systems; sensorized seat

---

1. Introduction

Modern lifestyle is more and more characterized by the fact that people are moving less and are spending an increasing amount of time sitting [1–4]. This takes into account the time spent working in front of a PC or using means of transportation to move from one place to another. Many leisure activities are also performed while seated (e.g., reading a book, watching television).

A sedentary lifestyle is dangerous for many reasons. It may lead to the development of chronic diseases (e.g., cardiovascular) [5,6] and may also affect psychological health [7]. Another problem related to the increase of time spent sitting is that the position adopted on the chair is often incorrect, thus creating health problems such as back pain [6] and headaches [7].

In recent years, people have become more aware of these problems. This is witnessed by the commercialization of height-adjustable sit–stand desks, but also by the spread of many low-cost devices such as smart watches (e.g., Fitbit trackers, Apple watches) and applications (e.g., Stand Up! The Work Break Timer, Time Out) that push the users to perform micro-breaks every 20–30 min of continuous sitting. In this regard, Sitting Posture Monitoring Systems (SPMS) have been introduced in the community, with the objective of detecting and understanding the position of a person in the seated position.

SPMS can rely on different technologies, which span from the use of wearable sensors [8–13], to the use of cameras [14–16], and also to the use of optoelectronic systems [17]. In all the considered cases, the aim is to provide the user with a minimally intrusive experience, so that the SPMS does not interfere with the normal behavior.

Among the available solutions, a state-of-the-art solution is that of equipping chairs with sensors able to collect data on the posture of the user [18–21]: Ishac et al. in [18] presented a cushion to be used in the backrest of the chair. The cushion is equipped with a pressure sensing array that allows the measurement of the pressure at 9 different points. The lack of information about the pressure on the seat reduces the amount of information that can be extracted.

Zemp et al. in [19] used an innovative approach based on machine learning to classify the data collected by 16 pressure sensors located on different parts of a chair (armrest, seat and backrest). While allowing an improved posture classification, the computational complexity of this algorithm is quite high.

Similarly, Roh et al. in [20] inserted load cells in the frame of the chair to record data that are processed through a machine learning approach. In this work, even if a dynamic evaluation of the sitting posture is achieved, data are used only to classify postures and not to evaluate the dynamic behavior of a subject in different conditions.

In a previous work [21], we equipped an office chair with textile sensors that allowed us to recognize 8 different sitting positions. We experimented a limitation since the smart chair allows only the detection of specific positions and not to monitor their evolution in time.

To evaluate the dynamic features of the seated position, we took inspiration from the use of force plates in posturography: in that domain, a three-component force plate is generally used to record over time where the Center of Pressure (COP) is located, by measuring and combining the vertical force component and the two torque components applied on it [22]. The COP is defined as the application point of the ground reaction force, and, in upright stance, it sways dynamically around an equilibrium point even in absence of external disturbances [23], a phenomenon that is generally denoted as postural sway. Postural sway is controlled by the Central Nervous System (CNS) in an autonomous way, and it is strictly related to the human capability of standing [24]. Human standing position is influenced by many disturbing factors (external, such as audio or visual stimuli, or internal, such as respiration or cardiac activity [25,26]) that are continuously compensated by the CNS.

Similarly, while seated, the human body is subject to a number of disturbances and, also in this condition, the CNS modulates the activity of muscles to maintain the equilibrium of the upper body segments (e.g., upper trunk, head, upper limbs) [27]. Correspondingly, in a seated position it is possible to determine where the Upper body Center of Pressure (UCOP) is located over a period of time, in order to evaluate, for example, if, in the presence of different external stimuli, or during task execution, a different way of swinging can be identified and to what it can be associated.

In both cases the equilibrium can be affected by different cognitive load levels of the task that a subject is performing. Different levels of cognitive load can be associated with variations of the electroencephalogram (EEG) signal [28]. In [29] the authors show that the EEG signal can be used to monitor different levels of stress. Moreover, it is known that an increase of cognitive load induces increasing levels of stress [30].

Among the techniques used to increase the cognitive load, the analysis of the EEG signals has proven that the Stroop Test activates brain areas related to attention [31]. Since this test can be a valuable tool to induce different cognitive loads on subjects, it has been used in the experiments presented in this work and it will be detailed in the following.

Anyway, EEG-based techniques require complex setup to record this type of signals. For this reason, EEG is not the most convenient means to monitor the level of stress in common daily life activities, such as in workplaces. Among the possible alternatives, it has been demonstrated that both eye movements and postural sway can be associated to different levels of cognitive involvement: in [32], 16 volunteers were asked to perform visual tasks while standing and they highlight the presence of a synergic relation between recorded eye and body movements for high precision tasks; moreover, even if some categories of trained people (e.g., activities that requires more equilibrium such as dancing) are in general more stable, everybody has a tendency to swing more when a demanding task is required [33], and this evidence makes space for the possibility to monitor cognitive load through the variations in postural sway. Following along this line, in [34], a Nintendo Wii balance board was selected to analyze the postural sway of older adults with no cognitive disability while performing predefined tasks in a home environment. In this case also, postural sway has been correlated to the cognitive status of the examined subjects.

In this work, the seated postural sway was analyzed using an innovative instrumented chair able to collect in a non-invasive way the instantaneous position of the UCOP. To test its validity, we asked subjects to perform a task sequence with increasing levels of cognitive engagement. As demonstrated in [21], increasing levels of cognitive engagement modify the posture that, starting from a relaxed seated condition will reach a stressed one, but up to now this was not quantified by the amount of seated postural sway associated with the observed modifications in posture.

2. Materials and Methods

2.1. Instrumented Chair Design

To evaluate the seated postural sway, a new instrumented chair has been designed and realized starting from previous experiences [21,35,36], where pressure sensors have been used to determine static postures of the users. In this work, an office chair was equipped with a set of four force sensors placed under its Load Plane (LP, i.e., where the pelvis loads up the chair). This set up was used to extract the instantaneous position of the UCOP mediolateral and anteroposterior coordinates (i.e., X_{UCOP} and Y_{UCOP}), under the hypothesis that the remaining body segments (thighs, legs, foot) do not concur to loading the seat.

In analogy to posturography in an upright stance, the required dynamic components to calculate the UCOP coordinates are: the resultant vertical force applied on LP F_z, that is the force component acting along the perpendicular Z axis; and the two torques M_x and M_y applied around the orthogonal axes identified on the load plane, identified as X and Y (Figure 1).

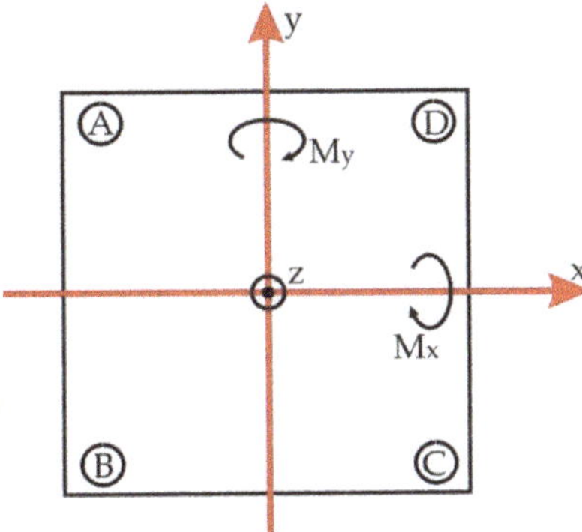

Figure 1. Scheme of the instrumented chair load plane: The reference system for the dynamic components are identified by (x,y,z) axes, while the load cells positions are identified by the {A,B,C,D} markers.

Considering that M_x and M_y are defined as the vector product of F_z and the distance of its point of application on LP from each axis, the values of the UCOP coordinates, X_{UCOP} and Y_{UCOP}, can be obtained as:

$$X_{UCOP} = \frac{M_y}{F_z};\ Y_{UCOP} = -\frac{M_x}{F_z}. \tag{1}$$

To evaluate these three dynamic components, four load cells can be inserted in the positions {A,B,C,D} of the LP (Figure 1) and the four force components $\{F_A, F_B, F_C, F_D\}$ can be measured. Consequently, F_z can be obtained as:

$$F_z = F_A + F_B + F_C + F_D. \tag{2}$$

Each load cell produces a voltage output V_P that in linear conditions results:

$$F_P = k_P V_P \tag{3}$$

where P = {A,B,C,D} represents the position and k_P is a proportionality constant.

If F_Z is applied in the center of the LP (i.e., $X_{UCOP} = 0$, $Y_{UCOP} = 0$) the four load cells are equally loaded and the torques applied around the X and Y axes are equal to zero, so it results:

$$k_A V_A = k_B V_B = k_C V_C = k_D V_D = 0.25 \times F_z. \tag{4}$$

Considering that four load cells with the same features are used (thus the proportionality constant is the same for all the load cells), the previous equation can be approximated by:

$$F_z = K_z(V_A + V_B + V_C + V_D) = K_z V_{Fz} \tag{5}$$

with K_Z the common constant of proportionality for the F_z measurement.

When F_Z is applied on a point that does not lie on one or both axes, one or both torque components are applied on the LP.

For example, when F_z is applied in a generic point on the x axis different from zero, a M_x component results on the LP; in this case the four load cells are loaded in a different way and it results that:

$$M_x = \frac{1}{2}[(k_B V_B + k_C V_C) - (k_A V_A + k_D V_D)] \tag{6}$$

where *l* is the distance between A and B (or between C and D).

This equation represents the sum of each torque contribution given by the force reactions, due to F_z solicitation, applied in the four P positions. Since the hypothesis of using load cells with the same features still holds, the previous equation can be approximated by:

$$M_x = \frac{lK_z}{2}[(V_B + V_C) - (V_A + V_D)] = K_x[(V_B + V_C) - (V_A + V_D)] = K_x V_{Mx} \tag{7}$$

with K_y a general constant of proportionality.

With similar considerations, it is possible to write that:

$$M_y = K_y[(V_C + V_D) - (V_A + V_B)] = K_y V_{My}. \tag{8}$$

It is then possible to measure F_z, M_X and M_y if the three parameters $\{K_x, K_y, K_z\}$ and the four voltage values $\{V_A, V_B, V_C, V_D\}$ are known. This computation allows to compute the UCOP coordinates, X_{UCOP} and Y_{UCOP}.

Each force component can be measured by a load cell and, after the signals are acquired, the mechanical quantities values can be calculated combining the digitalized data. These equations are valid in ideal conditions, where the cross-talk effect among channels (i.e., the influence of a specific load on all output channels, given by multiple factors) is neglected. In real conditions, every mechanical quantity is obtained taking in account cross-talk effects and properly modifying the given equations with a corrective factor; this is usually achieved by a calibration procedure that will be described in the following.

Using a normal office chair, four commercial load cells were placed in the four corners of its seat, to realize an instrumented system, where an LP can be identified, with the features described above. A custom frame was realized and used to partially replace the chair frame to obtain four areas of where to place the load cells (Figure 2). The custom frame was designed in order not to flex during dynamic solicitation, and not to affect the measures performed with the load cells.

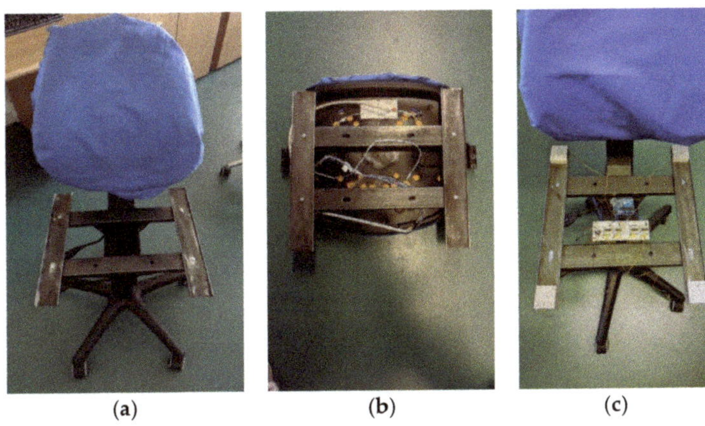

Figure 2. Custom frame realized to house the load cells: (**a**) first part of the new frame fixed on the chair structure; (**b**) second part of the new frame fixed under the seat; (**c**) load cells housing.

The load cells used in this project are the FX1901 class (Measurement Specialties Inc., TE Connectivity Group, USA), while the specific used model (FX1901-0001-0100-L) is a 1% load cell device with full scale ranges of 100lb compression (about 45kg). These devices, specifically designed for force sensing in "smart" consumer and medical products, use micro-machined piezoresistive strain gauges, showing a ratiometric span of 20 mV/V.

To place the load cells under the LP, a specific adaptor was designed and realized by means of a 3D printing process. A Stratasys Objet30 Prime was used: this machine allows polyJet 3D printing that allows the realization of high-resolution components by jetting microscopic layers of liquid photopolymer onto a build tray and instantly curing them with UV light. This accuracy in realization, comparable with a high accuracy machining of metal alloy, was necessary to couple the 4 load cells with the flat surface of the metal frame where they have been housed. In Figure 3, one of the realized adaptors is shown: the load cell is fixed to the lower part of the adaptor and is covered by a two-piece top part. This part presents a cylindrical element that is in contact with the load-sensitive part of the load cell and with the top part of the custom chair frame, thus transmitting directly to the load cell the force applied on the LP.

Each load cell, as mentioned above, was realized using strain gauges that are assembled in a full Wheatstone bridge configuration. As mentioned in its datasheet, a 4-wire connection is provided: 2 wires are used to supply the Wheatstone bridge $\{+V_{in}, -V_{in}\}$, while 2 wires are used to measure its outputs $\{+V_{out}, -V_{out}\}$.

The four load cells were driven independently using four instrumentation amplifiers, INA125 (Texas Instruments inc., USA). It is a high-accuracy instrumentation amplifier with a precision voltage reference used to provide a complete bridge excitation and that can receive and amplify differentially the bridge output. The gain was set using the dedicated single external resistor pin, using the datasheet information (selectable gain from 4 to 10,000) in order to best fit the voltage range of the adopted A/D converter. This last operation was performed using an Arduino Uno (Arduino cc) board, that through an ATmega328P microcontroller provided the necessary function to acquire signals. A specific firmware was realized to acquire from four input channels of the Arduino board the four INA125 output analog signals with a 100 Hz sampling frequency.

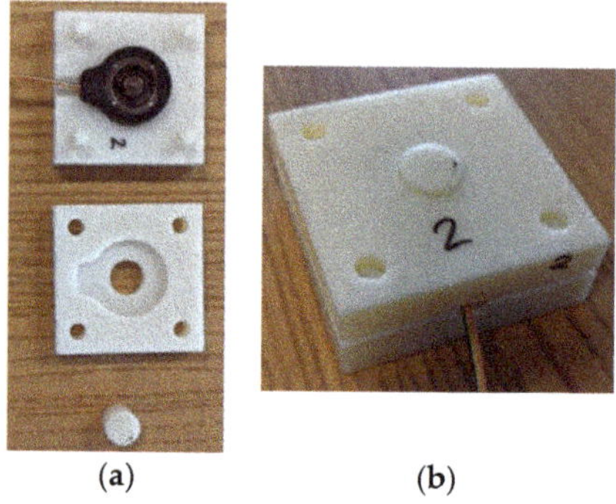

(a) (b)

Figure 3. Load cells housing adaptors: (**a**) open adaptors; (**b**) closed adaptor with enclosed load cell: the central cylindrical element is in contact with the load-sensitive part of the load cell.

The AD converter was managed by using a custom interface realized with Labview (National Instruments, USA) that allows the connection with the Arduino board via USB. The designed and realized Labview panel (i.e., Virtual Instrument, VI), allows the serial port connection to receive the digitalized data that are converted from bit levels to mechanical quantities using proper calibration coefficients. Moreover, the VI calculates in real time the UCOP coordinates and all data (i.e., F_z, M_x and M_y together with X_{UCOP} Y_{UCOP}) are saved in a file for further elaboration.

The mechanical quantities in real conditions, where cross-talk effects are present, are obtained from the acquired voltage values according to the following equations:

$$\begin{cases} F_z = a_{11}V_{Fz} + a_{12}V_{Mx} + a_{13}V_{My} \\ M_x = a_{21}V_{Fz} + a_{22}V_{Mx} + a_{23}V_{My} \\ M_y = a_{31}V_{Fz} + a_{32}V_{Mx} + a_{33}V_{My} \end{cases} \Rightarrow \begin{pmatrix} F_z \\ M_x \\ M_y \end{pmatrix} = \begin{pmatrix} a_{11} & a_{12} & a_{13} \\ a_{21} & a_{22} & a_{23} \\ a_{31} & a_{32} & a_{33} \end{pmatrix} \begin{pmatrix} V_{Fz} \\ V_{Mx} \\ V_{My} \end{pmatrix} \quad (9)$$

with $\begin{pmatrix} a_{11} \\ a_{22} \\ a_{33} \end{pmatrix} = \begin{pmatrix} K_z \\ K_x \\ K_y \end{pmatrix}$ and a_{ij} a generic calibration coefficient, constituting the calibration matrix, that is used to take into account the cross-talk effect. Since load cells' primary transducers work in their linear field, the calibration coefficients can be considered constant in the measurement range of the load cells.

Calibration coefficients have been obtained by positioning a graduate flat surface on the chair LP (Figure 4). The graduate surface has specific reference points that allow the application of controlled loads on the chair LP; as an example, by using known loads it is possible to apply only the F_z component or specific M_x and M_y, since the equidistant orthogonal axes from load cells couples are identified. Thus, considering three different load conditions, and reading the correspondent voltage output, it is possible to apply three times Equations (9) to obtain the calibration coefficients.

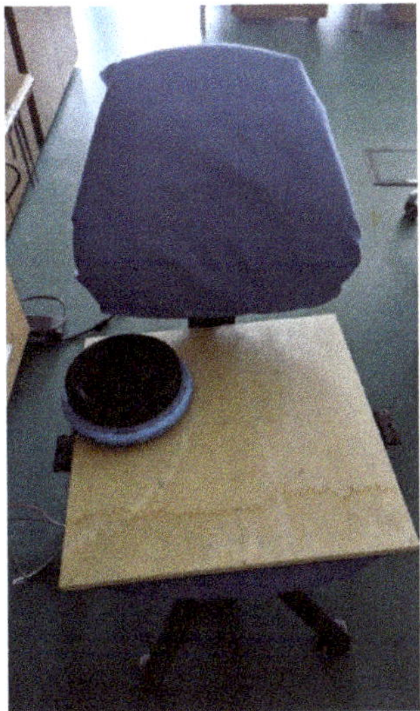

Figure 4. Load cells calibration setup performed using a graduate reference surface positioned on the chair LP and known loads.

2.2. Experimental Protocol

An experimental campaign has been designed and performed to extract the UCOP coordinates while asking the volunteers seated on the chair to undergo a task sequence with increasing levels of cognitive engagement defined through a dedicated test (i.e., Stroop test).

The Stroop test has been successfully employed for driving volunteers from a natural and relaxed position on the seat of a chair to different ones associated with a stressing condition while increasing the difficulty level of the performed task. Details of the designed Stroop test are given in the following.

2.2.1. Stroop Test

In the Strop test, a stress condition in the participants is induced, by asking them to perform a series of similar cognitive tasks with an increasing level of difficulty and engagement [37]. In previous works, it has been demonstrated that the increasing engagement in such kinds of tests gives rise to a different level of stress in the participant [38] and as a consequence, a different sitting posture on the chair is taken [21].

The test used in this study is based on displaying some words on a screen that the participant has to read in a predefined period of time: in particular, the selected words are the name of a color (e.g., "blue", "red", "black", etc.) and have been displayed written in black and white in a first phase while in a colored form in a second phase of the test. To best achieve a variation of the cognitive engagement and to increase the level of stress induced in the task, the phases are planned in order to induce the Stroop effect: in the first phase of the test, the participant is instructed through written indications to read the words displayed on the screen, while in the second phase he/she is invited to read aloud the color of the word, regardless of the meaning of the word (this will be called PHA in the following). Then the combinations (color–meaning) in the successive phases are: the name and the color are matching (e.g., the word "blue" is displayed using a blue color, PHB in the following); and the name and the color are not matching (e.g., the word "blue" is displayed using a red color, PHC in the following). Different sets of words are displayed in sequence, giving a fixed time slot to the participant to complete the reading of each set. In PHB, the task is very easy since the color of the word corresponds to its meaning. In PHC, the task is more demanding, and the number of mistakes made by the participant increases (e.g., not all words in the set are read or the meaning of the word is read instead of its color). This is mainly due to the time provided to the participants to complete the task being limited. Moreover, the participant is warned of the errors he/she made, and this induces a stressing condition that usually is maintained and growing till the end of the test.

In this work, a Stroop test consisting of 25 PowerPoint slides was realized and it was presented to a population of volunteers while sitting on the designed chair. In particular, according to the results obtained in [21], the test sequence (PHA, PHB, PHC) is described in the following and shown in Figure 5:

(a) PHA: 6 sets of black and white words representing color names, that were included in the test to make the participant familiar with the required task (i.e., the readings of words on a display in a predefined time slot) (Figure 5a).
(b) PHB: 6 sets of colored words representing color names with a correspondence between the adopted color and the meaning (Figure 5b).
(c) PHC: 12 sets of colored words representing color names with mismatching color and meaning (Figure 5c).

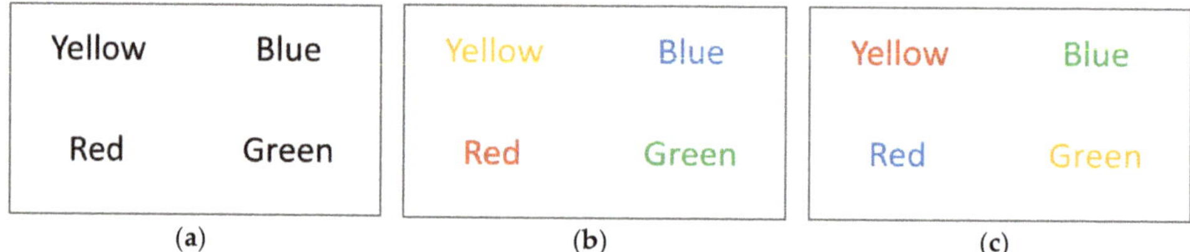

Figure 5. Black and white color words (**a**); matching color words (**b**); and non-matching color words (**c**) displayed respectively during phases PHA, PHB and PHC of the Stroop test.

Within each phase, the word set presented to the volunteers has an increasing number of words composing a slide that has been displayed in full screen mode. A fixed time delay for transition between consecutive sets has been set according to [21] to be able to induce the Stroop effect. Moreover, this choice allows for uniform test conditions for the whole population of participants, together with the standardization of the adopted font size (96 pt.) for all the displayed words.

The adopted settings are reported in Table 1.

Table 1. Schedule of the Stroop test—in the first phase the words are in black and white, in the second phase the color of words matches the word meaning, and in the third phase the color of words does not match the word meaning.

	Number of Words Per Set	Number of Slides	Time Per Slide [s]
PHA: Black and white words	3	2	2
	4	2	2.5
	5	1	3
	6	1	3.5
PHB: Colored words with color matching the meaning	3	2	2
	4	2	2.5
	5	1	3
	6	1	3.5
PHC: Colored words with color not matching the meaning	3	2	2.5
	4	2	3
	5	2	3.5
	6	2	4
	9	2	6
	12	2	8

2.2.2. Experimental Protocol

The tests have been presented to a group of 95 participants (61 males and 34 females), with age of 29 ± 12 years (mean ± SD). All participants filled a form declaring their agreement to participate in this scientific test. All volunteers had no visual impairments (i.e., none of them wore glasses or contact lenses, all volunteers were checked for color blindness) and in general, were healthy with no apparent or declared mobility problems (e.g., use of walking aids). Since the sets of words were written in Italian, all the selected participants were native Italian speakers. The participants were recruited among the population visiting the "Maker Faire 2019" (The European edition, Rome, Italy).

Since the chair was equipped only with force sensors on the seat, the backrest was removed, so the chair was used like a stool in its use during the test. Each participant was invited to sit in a natural position, without crossing their legs. Participants were not informed about the scope of the test before conducting it, in order not to influence their behavior during its execution. Participants were briefly instructed about the tasks to be conducted, and further detailed instructions (i.e., "read the word on the slide" or "read the color of the words") were provided during the displaying of the words sets (i.e., at the beginning and at each phase transition of the test). The participants were asked to read aloud as quick as possible the words' meaning for the sets displayed in PHA and PHB, and the color of each word regardless of its meaning for the sets displayed in PHC. After the conclusion of the test, each participant was informed about the presence of the sensors under the seat and about the recorded data. Moreover, participants were informed that they were part of a scientific study, and everybody confirmed the given consensus to be part of the studied population. Nobody felt the presence of sensors during the test or a feeling different than that of sitting on a normal stool. Recorded data were stored in a database, together with the personal information of the volunteers (i.e., age and gender), and associated to a code for each participant. A list matching the participant name and the code was encrypted and stored in a different storage device from that of the database.

2.3. *Data Analysis*

From the acquired and calibrated dynamic data (i.e., F_z, M_x and M_y), the UCOP coordinates, X_{UCOP} and Y_{UCOP}, with respect to the chair origin were calculated for each phase according to Equation (1), removed of their average value for each phase, and low-pass filtered through a third-order zero phase Butterworth digital filter with a 20-Hz cut-off frequency. This produced, for each phase, a 2D time series, $\{X_{UCOP}A, Y_{UCOP}A\}$, $\{X_{UCOP}B, Y_{UCOP}B\}$ and $\{X_{UCOP}C, Y_{UCOP}C\}$, respectively. Then, the corresponding resulting distances RDA, RDB and RDC (i.e., the magnitude of the displacement of the instantaneous position of UCOP from its mean value) were calculated. From these time series, in agreement with the definitions given in [22], mean distance values (MD, MDX and MDY), RMS distance values (RDIST, RDISTX and RDISTY), range values (RANGE, RANGEX and RANGEY), Mean Velocity values (MVELO, MVELOX and MVELOY), the 95% confidence circle area value (AREA-CC), the 95% confidence ellipse area value (AREA-CE) and the sway area value (SWAREA) were calculated for every subject and for each phase. Moreover, for each parameter and for each phase, the mean value and the standard deviation were calculated for all subjects. Pairwise t-tests were conducted with phases as a grouping factor for all the calculated parameters. In order to verify whether there was a significant difference in parameters among the three phases, a p-value < 0.05 was chosen, but also different acceptable values of p (i.e., $p < 0.01$, $p < 0.001$ and $p < 0.0001$) were selected to indicate different levels of statistical effect.

3. Results

The time trends of X_{UCOP}, Y_{UCOP} and RD for one of the participants are displayed in Figure 6, together with the plots of the corresponding stabilograms (i.e., X_{UCOP}-vs-Y_{UCOP} plot). Different colors are used in the figures to highlight differences among the three phases (green for PHA, red for PHB, and black for PHC).

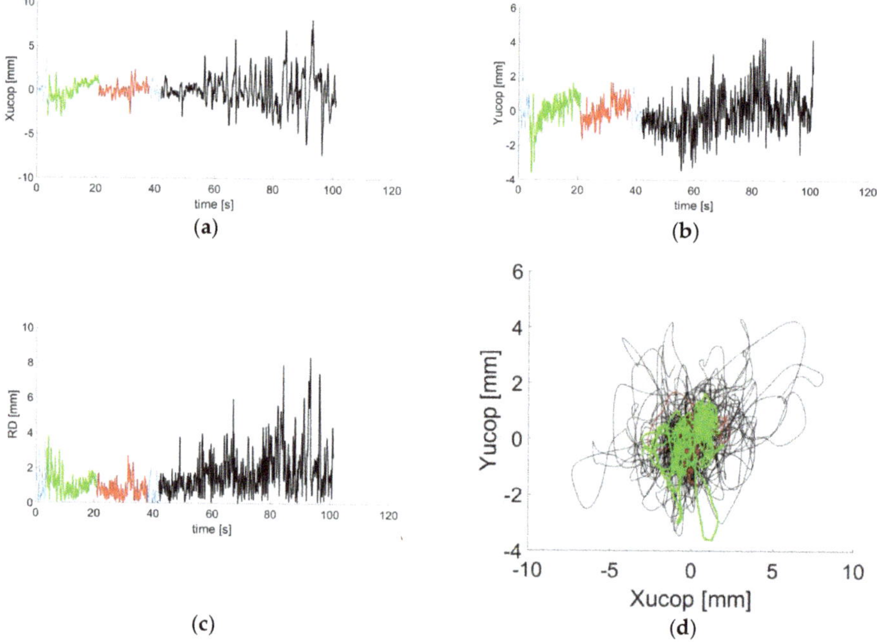

Figure 6. Time trends of the XUCOP (**a**), YUCOP (**b**) and RD (**c**), together with the corresponding stabilogram (**d**) for one of the participants.

In the displayed time series, the increase of the X_{UCOP} and Y_{UCOP} range, together with the calculated RD, is remarkable throughout the test from the least engaging part (PHA) to the most demanding one (PHC). In addition, the stabilogram plot shows a significant increase in the swept area in PHC as compared to PHA and PHB. This increment of the seated sway excursion demonstrates that when volunteers are involved in a demanding cognitive task, there is a lower stability on the seat with respect to the phases of the test where the task to be performed is simpler. This can be due to the growing of a stressing condition that can cause a different position on the chair, as demonstrated in [21], and can also induce a different dynamic behavior.

The results of the statistical analysis are shown in Table 2, together with the mean values of the calculated parameters for all participants, and the corresponding standard deviations. For all parameters, except the sway area, the difference between coupled distributions resulted highly significant for PHA vs. PHC and PHB vs. PHC, thus validating the hypothesis that a different strategy for controlling the sway is assumed by subjects when cognitive engagement increases. Moreover, for the range parameters (i.e., RANGE, RANGEX and RANGEY), the difference between PHA and PHB resulted significant: the difference is higher for the y coordinate (i.e., anteroposterior) and this can be explained by the biomechanics of the hip. During a seated position, this joint has reduced swinging along the mediolateral direction (i.e., x direction) with respect to the anteroposterior one (i.e., y direction), so when posture is not well controlled, a greater instability along y is found.

Table 2. UCOP-based descriptive parameters computed for all the volunteers: for each parameter the mean and std values calculated over all participants have been reported for each of the three phases of the test. For the same parameters, the p-value obtained from the multiple pairwise t-tests between phases have been calculated and different level of significance have been assessed as reported in the following: * $p < 0.05$, ** $p < 0.01$, **** $p < 0.0001$.

Measure	PHA		PHB		PHC		*p*-values					
	Mean	std	Mean	std	Mean	std	PHA-PHB	PHA-PHC	PHB-PHC	PHA-PHB	PHA-PHC	PHB-PHC
mean distance (mm)	**1.95**	**1.30**	**1.66**	**1.44**	**4.72**	**4.14**	**0.1481**	$\mathbf{1.06 \times 10^{-15}}$	$\mathbf{1.22 \times 10^{-10}}$		****	****
mean distance x (mm)	1.16	0.74	1.07	1.06	2.84	2.83	0.5376	2.90×10^{-12}	4.68×10^{-8}		****	****
mean distance y (mm)	1.30	1.05	1.03	0.93	3.14	2.85	0.0628	4.21×10^{-15}	9.41×10^{-11}		****	****
rms distance (mm)	**2.45**	**1.67**	**1.96**	**1.69**	**5.77**	**5.03**	**0.0457**	$\mathbf{6.38 \times 10^{-16}}$	$\mathbf{4.18 \times 10^{-11}}$	*	****	****
rms distance x (mm)	1.59	1.03	1.40	1.35	3.81	3.61	0.2878	2.50×10^{-13}	6.02×10^{-9}		****	****
rms distance y (mm)	1.77	1.44	1.30	1.10	4.14	3.74	0.0124	2.18×10^{-15}	2.66×10^{-11}	*	****	****
range (mm)	**13.11**	**9.57**	**9.70**	**7.93**	**37.54**	**33.40**	**0.0082**	$\mathbf{7.41 \times 10^{-20}}$	$\mathbf{2.25 \times 10^{-13}}$	**	****	****
range x (mm)	10.25	7.17	8.14	7.12	28.72	27.62	0.0435	9.16×10^{-17}	3.69×10^{-11}	*	****	****
range y (mm)	10.72	8.82	7.30	5.20	29.89	26.80	0.0014	9.68×10^{-20}	8.33×10^{-14}	**	****	****
mean velocity (mm/s)	**11.33**	**5.47**	**10.78**	**5.84**	**4.72**	**2.97**	**0.5044**	$\mathbf{3.98 \times 10^{-20}}$	$\mathbf{2.32 \times 10^{-16}}$		****	****
mean velocity x (mm/s)	6.86	3.31	6.44	3.72	2.79	1.84	0.4176	1.55×10^{-19}	3.48×10^{-15}		****	****
mean velocity y (mm/s)	7.48	3.92	7.22	4.30	3.18	2.07	0.6617	3.81×10^{-17}	2.80×10^{-14}		****	****
95% confidence circle area (mm^2)	**6.19**	**2.11**	**5.35**	**2.09**	**9.20**	**3.86**	**0.0067**	$\mathbf{4.54 \times 10^{-19}}$	$\mathbf{4.25 \times 10^{-15}}$	**	****	****
95% confidence ellipse area (mm^2)	**66.36**	**94.90**	**47.43**	**79.17**	**433.16**	**732.97**	**0.1371**	$\mathbf{1.70 \times 10^{-10}}$	**8.27E-07**		****	****
sway area (mm^2/s)	**10.36**	**13.32**	**7.45**	**10.19**	**11.75**	**18.59**	**0.0932**	**0.1145**	**0.0497**			*

Finally, significant differences can be observed in the RMS distance values (RDIST and RDISTY) and in the 95% confidence circle area (mm^2).

Figures 7–11 show the graphical representation of the calculated parameters' distributions. For MD values, it is noticeable that the increased engagement in PHC for the requested task significantly increases the distance from the UCOP center in both sway directions (Figure 7). The same results can be observed considering the RMS distances values (Figure 8). For both parameters, a greater excursion in the y direction (i.e., anteroposterior) is noticeable.

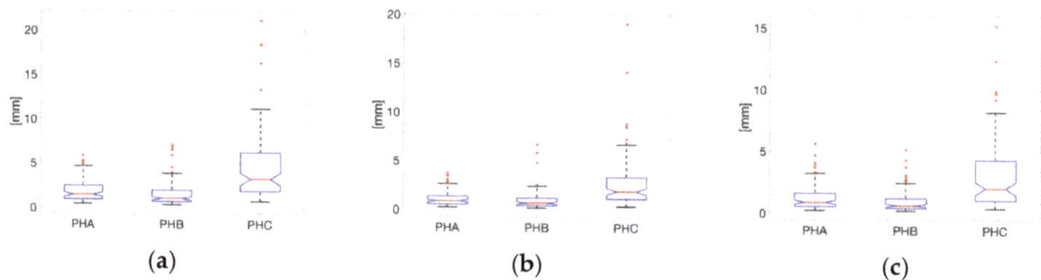

Figure 7. Mean distance distributions for each of the three phases PHA, PHB and PHC: (**a**) MD distributions, (**b**) MDX distributions and (**c**) MDY distributions.

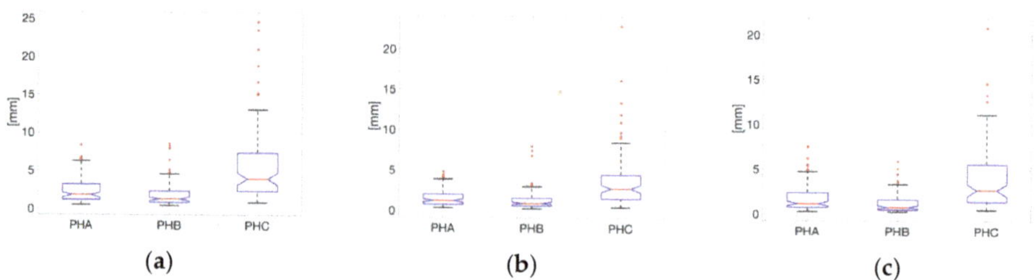

Figure 8. RMS distance distributions for each of the three phases PHA, PHB and PHC: (**a**) RDIST distributions, (**b**) RDISTX distributions and (**c**) RDISTY distributions.

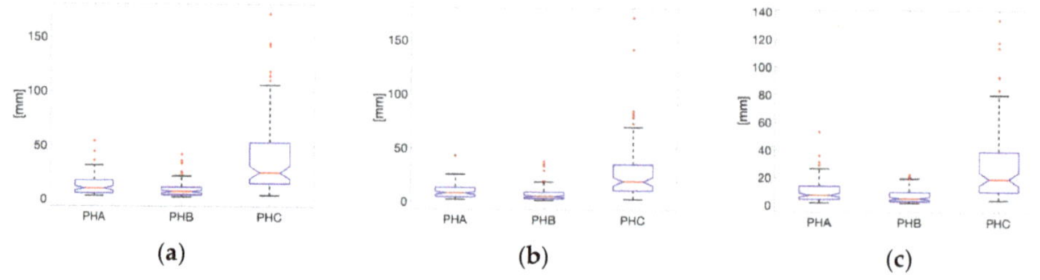

Figure 9. Range distributions for each of the three phases PHA, PHB and PHC: (**a**) RANGE distributions, (**b**) RANGEX distributions and (**c**) RANGEY distributions.

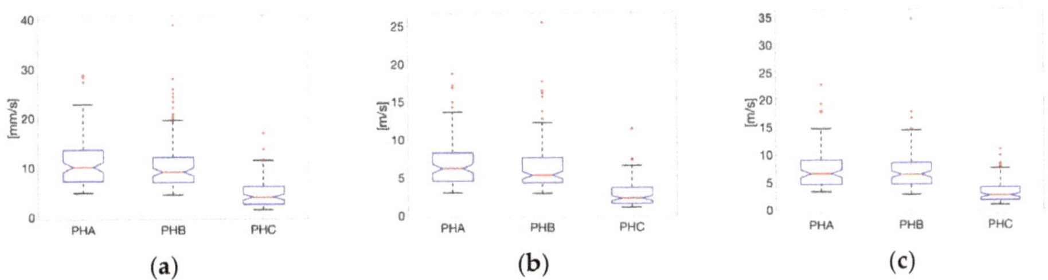

Figure 10. Mean velocity distributions for each of the three phases PHA, PHB and PHC: (**a**) MVELO distributions, (**b**) MVELOX distributions and (**c**) MVELOY distributions.

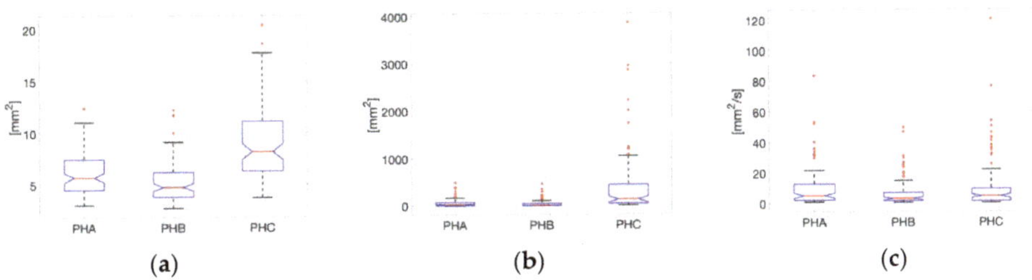

Figure 11. Area Parameters distributions for each of the three phases PHA, PHB and PHC: (**a**) AREA-CC distributions, (**b**) AREA-CE distributions and (**c**) SWAREA distributions.

These results are confirmed considering the range parameters distributions (Figure 9), where higher values of the maximum and minimum coordinates result in the UCOP excursion during PHC. It is also important to note that the range values in PHB are significantly lower than the ones in PHA: it can be speculated that after the first phase of the test, the volunteers reach a stable position where they feel comfortable both with the sitting posture and in performing the task. This condition is strongly modified in PHC and, since the total duration of the test is chosen not to induce muscular fatigue, the change of strategy in swaying can be explained by the increase in the difficulty of the task that induces a higher cognitive engagement.

Considering the velocity parameters distributions (Figure 10), a significant decrease in the velocity of the UCOP trajectory appeared in PHC, compared to both PHA and PHB. This evidence, together with the assessed high excursion, can confirm a decreased capability in controlling equilibrium in the most demanding condition: in fact, if the greater excursion would have been caused by an active rapid movement, the velocity would not have decreased as shown. The decrease of these values can therefore be explained by a reduced capability in controlling the UCOP trajectory and "corrective" actions performed to remain in a seated equilibrium.

Finally, area parameters distributions (Figure 11) reveal a larger area swept by the UCOP trajectory and a general instability achieved by subjects in PHC than in PHA and PHB, especially considering the 95% confidence circle and ellipse areas.

4. Conclusions

In this paper, a new system to assess if engaging cognitive activities influence the postural sway of a subject seated on a chair is presented. The designed and realized system, based on a chair equipped with force sensors that allows the dynamics description of the examined motor act, provides the time trend of the volunteers' position. These data can provide useful information on the adopted strategies in the seated postural control in subjects directed towards a complex cognitive task in a reduced amount of time. The obtained results show a significant difference in sway strategy and a coherent variation for a high number of examined volunteers, thus supporting the hypothesis that the higher demand the cognitive task has, the lower the stability in the seated posture. The realized system is based on the integration of sensors on a normal office chair, whose physical features are not modified. Moreover, the system is designed to be powered with batteries and can be improved by embedding a wireless transmission. All these features allow the use of the chair in a wide range of situations, such as a smart working environment, where the device can provide useful information about the risk related to working activities due to the task-induced cognitive load.

As a future development, the system could be further validated by using additional techniques for cognitive load assessment, such as EEG analysis, testing the same setup in conditions similar to the ones presented in this work.

The realized device will be completed with seat backrest sensing, in order to obtain a system that allows a complete analysis of the adopted posture. This can be very useful to monitor activities that require a prolonged seated time, and where the cognitive engagement is constant: in this case, the postural sway could be a useful indicator of global fatigue, thus giving important information on the correct time schedule of the performed working tasks and to prevent all the pathologies related to the adoption of improper seated postures.

Author Contributions: Conceptualization, D.B. and F.B.; methodology, D.B., S.C., and F.B.; software, D.B. and F.B.; validation, D.B., and S.C.; formal analysis, D.B. and F.B.; investigation D.B., M.S., and F.B.; resources, S.C. and M.S.; writing—original draft preparation, D.B. and F.B.; writing—review and editing, D.B., S.C., M.S. and F.B. All authors have read and agreed to the published version of the manuscript.

Funding: This research received no external funding.

Acknowledgments: The authors would like to thank Moses Mariajoseph, Giorgio Scordino and Arturo Zezza for their contribution in recruiting the volunteers for the experimental campaign.

Conflicts of Interest: The authors declare no conflict of interest.

References

1. Mielke, G.I.; Burton, N.W.; Turrellc, G.; Browna, W.J. Temporal trends in sitting time by domain in a cohort of mid-age Australian men and women. *Maturitas* **2018**, *116*, 108–115. [CrossRef]
2. Medina, C.; Tolentino-Mayo, L.; López-Ridaura, R.; Barquera, S. Evidence of Increasing Sedentarism in Mexico City during the Last Decade: Sitting Time Prevalence, Trends, and Associations with Obesity and Diabetes. *PLoS ONE* **2017**, *12*, e0188518. [CrossRef] [PubMed]
3. Owen, N.; Sparling, P.B.; Healy, G.N.; Dunstan, D.W.; Matthews, C.E. Sedentary Behavior: Emerging Evidence for a New Health Risk. *Mayo Clin. Proc.* **2010**, *85*, 1138–1141. [CrossRef] [PubMed]
4. Loyen, A.; Chey, T.; Engelen, L.; Bauman, A.; Lakerveld, J.; van der Ploeg, H.P.; Brug, J.; Chau, J.Y. Recent Trends in Population Levels and Correlates of Occupational and Leisure Sitting Time in Full-Time Employed Australian Adults. *PLoS ONE* **2018**, *13*, e0195177. [CrossRef] [PubMed]
5. Ekelund, U.; Steene-Johannessen, J.; Brown, W.J.; Fagerland, M.W.; Owen, N.; Powell, K.E.; Bauman, A.; Lee, I.M. Lancet Physical Activity Series 2 Executive Committe; Lancet Sedentary Behaviour Working Group. Does physical activity attenuate, or even eliminate, the detrimental association of sitting time with mortality? A harmonised meta-analysis of data from more than 1 million men and women. *Lancet* **2016**, *388*, 1302–1310. [CrossRef]
6. Ikegami, K.; Hirata, H.; Ishihara, H.; Guo, S. Development of a Chair Preventing Low Back Pain with Sitting Person Doing Hand Working at the Same Time. In Proceedings of the 2018 IEEE International Conference on Mechatronics and Automation (ICMA), Changchun, China, 5–8 August 2018; pp. 1760–1764. [CrossRef]
7. Kim, J.B.; Yoo, J.K.; Yu, S. Neck–Tongue Syndrome Precipitated by Prolonged Poor Sitting Posture. *Neurol. Sci.* **2014**, *35*, 121–122. [CrossRef]
8. Sardini, E.; Serpelloni, M.; Pasqui, V. Daylong sitting posture measurement with a new wearable system for at home body movement monitoring. In Proceedings of the 2015 IEEE International Instrumentation and Measurement Technology Conference (I2MTC), Pisa, Italy, 11–14 May 2015; pp. 652–657. [CrossRef]
9. Dunne, L.; Walsh, P.; Smyth, B.; Caulfield, B. Design and Evaluation of a Wearable Optical Sensor for Monitoring Seated Spinal Posture. In Proceedings of the 2006 10th IEEE International Symposium on Wearable Computers, Montreux, Switzerland, 11–14 October 2006; pp. 65–68.

10. Fida, B.; Bernabucci, I.; Bibbo, D.; Conforto, S.; Proto, A.; Schmid, M. The Effect of Window Length on the Classification of Dynamic Activities through a Single Accelerometer, Biomedical Engineering. In Proceedings of the IASTED International Conference Biomedical Engineering (BioMed), Zurich, Switzerland, 23–25 June 2014.
11. Fida, B.; Bibbo, D.; Bernabucci, I.; Proto, A.; Conforto, S.; Schmid, M. Real Time Event-Based Segmentation to Classify Locomotion Activities through a Single Inertial Sensor. In Proceedings of the 5th EAI International Conference on Wireless Mobile Communication and Healthcare, London, UK, 14–16 October 2015.
12. Yamato, Y. Experiments of Posture Estimation on Vehicles Using Wearable Acceleration Sensors. In Proceedings of the 2017 IEEE 3rd International Conference on Big Data Security on Cloud (BigDataSecurity), IEEE International Conference on High Performance and Smart Computing, (HPSC) and IEEE International Conference on Intelligent Data and Security (IDS), Beijing, China, 26–28 May 2017; pp. 14–17.
13. Massé, F.; Bourke, A.K.; Chardonnens, J.; Paraschiv-Ionescu, A.; Aminian, K. Suitability of commercial barometric pressure sensors to distinguish sitting and standing activities for wearable monitoring. *Med. Eng. Phys.* **2014**, *36*, 739–744. [CrossRef]
14. Estrada, J.; Vea, L. Sitting posture recognition for computer users using smartphones and a web camera. In Proceedings of the TENCON 2017 - 2017 IEEE Region 10 Conference, Penang, Malaysia, 5–8 November 2017; pp. 1520–1525. [CrossRef]
15. Amine Elforaici, M.E.; Chaaraoui, I.; Bouachir, W.; Ouakrim, Y.; Mezghani, N. Posture Recognition Using an RGB-D Camera: Exploring 3D Body Modeling and Deep Learning Approaches. In Proceedings of the 2018 IEEE Life Sciences Conference (LSC), Montreal, QC, Canada, 28–30 October 2018; pp. 69–72.
16. Wang, W.-J.; Chang, J.-W.; Haung, S.-F.; Wang, R.-J. Human Posture Recognition Based on Images Captured by the Kinect Sensor. *Int. J. Adv. Robot. Syst.* **2016**, *13*. [CrossRef]
17. D'Anna, C.; Varrecchia, T.; Bibbo, D.; Orsini, F.; Schmid, M.; Conforto, S. Effect of different smartphone uses on posture while seating and standing. In Proceedings of the 2018 IEEE International Symposium on Medical Measurements and Applications (MeMeA), Rome, Italy, 11–13 June 2018.
18. Ishac, K.; Suzuki, K. LifeChair: A Conductive Fabric Sensor-Based Smart Cushion for Actively Shaping Sitting Posture. *Sensors* **2018**, *18*, 2261. [CrossRef]
19. Zemp, R.; Tanadini, M.; Plüss, S.; Schnüriger, K.; Singh, N.B.; Taylor, W.R.; Lorenzetti, S. Application of Machine Learning Approaches for Classifying Sitting Posture Based on Force and Acceleration Sensors. *BioMed Res. Int.* **2016**, *2016*, 5978489. [CrossRef]
20. Roh, J.; Park, H.J.; Lee, K.J.; Hyeong, J.; Kim, S.; Lee, B. Sitting Posture Monitoring System Based on a Low-Cost Load Cell Using Machine Learning. *Sensors* **2018**, *18*, 208. [CrossRef] [PubMed]
21. Bibbo, D.; Carli, M.; Conforto, S.; Federica, B. A Sitting Posture Monitoring Instrument to Assess Different Levels of Cognitive Engagement. *Sensors* **2019**, *19*, 455. [CrossRef] [PubMed]
22. Prieto, T.E.; Myklebust, J.; Hoffmann, R.; Lovett, E.; Myklebust, B. Measures of postural steadiness: Differences between healthy young and elderly adults. *IEEE Trans. Biomed. Eng.* **1996**, *43*, 956–966. [CrossRef] [PubMed]
23. Błaszczyk, J.W. The use of force-plate posturography in the assessment of postural instability. *Gait Posture* **2016**, *44*, 1–6. [CrossRef] [PubMed]
24. Morasso, P.G.; Sanguineti, V. Ankle Muscle Stiffness Alone Cannot Stabilize Balance during Quiet Standing. *J. Neurophysiol.* **2002**, *88*, 2157–2162. [CrossRef] [PubMed]
25. Schmid, M.; Conforto, S.; Bibbo, D.; D'Alessio, T. Respiration and postural sway: Detection of phase synchronizations and interactions. *Hum. Mov. Sci.* **2004**, *23*, 105–119. [CrossRef]
26. Conforto, S.; Schmid, M.; Camomilla, V.; D'Alessio, T.; Cappozzo, A. Hemodynamics as a possible internal mechanical disturbance to balance. *Gait Posture* **2001**, *14*, 28–35. [CrossRef]
27. Goodworth, A.D.; Tetreault, K.; Lanman, J.; Klidonas, T.; Kim, S.; Saavedra, S. Sensorimotor control of the trunk in sitting sway referencing. *J. Neurophysiol.* **2018**, *120*, 37–52. [CrossRef]
28. Galin, D.; Johnstone, J.; Herron, J. Effects of task difficulty on EEG measures of cerebral engagement. *Neuropsychologia* **1978**, *16*, 461–472. [CrossRef]
29. Hou, X.; Liu, Y.; Sourina, O.; Tan, Y.R.E.; Wang, L.; Mueller-Wittig, W. EEG Based Stress Monitoring. In Proceedings of the 2015 IEEE International Conference on Systems, Man, and Cybernetics, Kowloon, China, 9–12 October 2015; pp. 3110–3115. [CrossRef]

30. Russ, S.; MacKenzie, R.K.; Morrison, I.; Morse, J.; Johnston, M.; Bell, C.; Patey, R. 0094 The impact of demand and cognitive load on stress and performance in 2nd year medical students: A simulation study. *BMJ Simul. Technol. Enhanc. Learn.* **2015**, *1*, A10. [CrossRef]
31. West, R.; Bell, M.A. Stroop color—word interference and electroencephalogram activation: Evidence for age-related decline of the anterior attention system. *Neuropsychology* **1997**, *11*, 421–427. [CrossRef] [PubMed]
32. Bonnet, C.T.; Davin, D.; Hoang, J.-Y.; Baudry, S. Relations between Eye Movement, Postural Sway and Cognitive Involvement in Unprecise and Precise Visual Tasks. *Neuroscience* **2018**, *416*, 177–189. [CrossRef] [PubMed]
33. Stins, J.F.; Michielsen, M.E.; Roerdink, M.; Beek, P.J. Sway regularity reflects attentional involvement in postural control: Effects of expertise, vision and cognition. *Gait Posture* **2009**, *30*, 106–109. [CrossRef] [PubMed]
34. Leach, J.M.; Mancini, M.; Kaye, J.A.; Haye, T.L.; Horak, F.B. Day-to-Day Variability of Postural Sway and Its Association With Cognitive Function in Older Adults: A Pilot Study. *Front. Aging Neurosci.* **2018**, *10*, 126. [CrossRef]
35. Bibbo, D.; Gabriele, S.; Scorza, A.; Schmid, M.; Sciuto, S.A.; Conforto, S. Strain gauges position optimization in designing custom load cells for sport gesture analysis. In Proceedings of the IEEE 20th International Conference on E-Health Networking, Applications and Services (Healthcom), Ostrava, Czech Republic, 17–20 September 2018; pp. 1–6.
36. Bibbo, D.; Gabriele, S.; Scorza, A.; Schmid, M.; Sciuto, S.A.; Conforto, S. A Novel Technique to Design and Optimize Performances of Custom Load Cells for Sport Gesture Analysis. *IRBM* **2019**, *40*, 201–210. [CrossRef]
37. Stroop, J.R. Studies of interference in serial verbal reactions. *J. Exp. Psychol.* **1935**, *18*, 643–662. [CrossRef]
38. Renaud, P.; Blondin, J.-P. The Stress of Stroop Performance: Physiological and Emotional Responses to Color–Word Interference, Task Pacing, and Pacing Speed. *Int. J. Psychophysiol.* **1997**, *27*, 87–97. [CrossRef]

Article

Rate-Invariant Modeling in Lie Algebra for Activity Recognition

Malek Boujebli [1,*], Hassen Drira [2,3], Makram Mestiri [1] and Imed Riadh Farah [1]

1 Software Engineering, Distributed Applications, Decision Systems and Intelligent Imaging Research Laboratory (RIADI), National School of Computer Science (ENSI), University of Manouba, Manouba 2010, Tunisia; mmestiri@gmail.com (M.M.); imed.riadh.farah@gmail.com (I.R.F.)

2 IMT Lille Douai, Institut Mines-Télécom, Center for Digital Systems, F-59000 Lille, France; hassen.drira@imt-lille-douai.fr

3 Univ. Lille, CNRS, Centrale Lille, Institut Mines-Télécom, UMR 9189-CRIStAL, F-59000 Lille, France

* Correspondence: mboujebli@gmail.com

Received: 21 September 2020; Accepted: 4 November 2020; Published: 10 November 2020

Abstract: Human activity recognition is one of the most challenging and active areas of research in the computer vision domain. However, designing automatic systems that are robust to significant variability due to object combinations and the high complexity of human motions are more challenging. In this paper, we propose to model the inter-frame rigid evolution of skeleton parts as the trajectory in the Lie group $SE(3) \times \ldots \times SE(3)$. The motion of the object is similarly modeled as an additional trajectory in the same manifold. The classification is performed based on a rate-invariant comparison of the resulting trajectories mapped to a vector space, the Lie algebra. Experimental results on three action and activity datasets show that the proposed method outperforms various state-of-the-art human activity recognition approaches.

Keywords: activity recognition; rate invariance; Lie group

1. Introduction

Human activity recognition has attracted many research groups in recent years due to its wide range of promising applications in different domains, like surveillance, video games, physical rehabilitation, etc. In order to develop systems for understanding human behavior, visual data form one of the most important cues compared to verbal or vocal communication data. Moreover, the introduction of low cost depth cameras with real-time capabilities, like the Microsoft Kinect, which provide in addition to the classical red-green-blue (RGB) image, a depth image, makes it possible to estimate in real time a 3D humanoid skeleton thanks to the work of Shotton et al. [1]. This type of data brings several advantages as it makes the background easy to remove and allows extracting and tracking the human body, thus capturing the human motion in each frame. Additionally, the 3D depth data are independent of the human appearance (texture), providing a more complete human silhouette relative to the silhouette information used in the past. Thus, new datasets with RGB-depth (RGBD) data have been collected, and many efforts have been made on human action recognition. However, human activity understanding is a more challenging problem due to the diversity and complexity of human behaviors, and less effort has been made by previous approaches. The interaction with objects creates an additional challenge for human activity recognition. Actually, during a human–object interaction scene, the hands may hold objects that are hardly detected or recognized due to heavy occlusions and appearance variations. The high level information of the objects is needed to recognize the human–object interaction. Taking a glance at the past skeleton-based human activity recognition approaches, we can distinguish two categories: the first family of approaches considers the skeleton data as body parts, and the second family considers them as a set of joints, as categorized by [2]. The scope of the paper is related to the first family of approaches that first consider the human skeleton as a connected set of rigid segments and either model the temporal evolution of individual body parts [3] or focus on connected pairs of body parts and model the temporal evolution of joint angles [4,5]. More recently, Vemulapalli et al. [2] proposed to model a skeleton by all the possible rigid transformations between its segments. In other words, for each skeleton, hundreds of rotations and translations (let us assume L) are computed between all the skeleton segments to yield L points on the special euclidean group $SE(3)$. The evolution of the skeleton along frames generates a trajectory in $SE(3)^L$. The trajectories are later on mapped to the Lie algebra (the tangent space on the identity point of the special euclidean group). The main limitation of this approach is the distortions caused by this mapping especially for points far from the identity element. The authors proposed an improvement of this method by the rolling-based approach in order to minimize the distortions in the tangent space (Lie algebra) in [6]. In this paper, we propose to investigate transformations of each skeleton part along frames in the Lie group and not within the same frame as [2,6]. Compared to [2] and [6], the proposed model represents three main advantages:

- For a skeleton with n joints, we manage a trajectory in $SE(3)^{n-1}$ compared to a trajectory in a much bigger space $SE(3)^{n\times(n-1)}$ as modeled in [2,6]. This makes the proposed approach faster than that of [2,6].
- The mapping into the tangent space on the identity element does not cause distortions in the proposed approach as the transformations are considered for the same body segment across frames, and thus, the resulting points on the special euclidean group are close to the identity element.
- We model the object within the human–object interaction and present results on datasets including human–object interaction, which was not the handled in [2,6].

The main contributions of this work are:

- We perform a spatio-temporal modeling of skeleton sequences as trajectories on the special euclidean group.
- The rigid transformations of the object are modeled as an additional trajectory in the same manifold, while in [7], only the joint-based approaches were proposed.
- An elastic metric of the trajectories is proposed to model the time independently of the execution rate.
- Exhaustive experiments and comparative studies are presented on three benchmarks: a benchmark for action without objects (MSR-Action dataset), a benchmark for actions with object interaction (SYSU3D Human-Object Interaction dataset), and a benchmark with a mixture of action and human–object interaction (MSR Daily Activity dataset).

The paper is organized as follow: We provide a brief review of the existing literature in Section 2 and discuss the spatio-temporal modeling in Section 3. Section 4 presents the rate invariance modeling and classification. We present our experimental results in Section 5 and conclude the paper in Section 6.

2. State-of-the-Art

Currently, the recognition of human activities has become more popular in the computer vision committee, and this interest is translated by many applications into more activities such as surveillance, video games, physical rehabilitation, etc. In this case, we can distinguish three emerging branches in the research on the recognition of activities: (1) depth-based representation, (2) skeleton-based representation, and (3) RGB-D-based development. In this section, we will go back to the existing work of recognizing human activities captured by depth cameras and describe in more detail the literature on the specific shared structures of learning for the recognition of activities.

2.1. Depth-Based Representation

In [8,9], descriptors previously designed for the deep RGB channel were generalized to describe the geometry of the shape and construct a depth-based representation. The limitation of the approach proposed in [8] is the sensitivity to the point of view as the sampling scheme depends on the view. Along similar lines, Oreifej and Liu [3] used the histogram of oriented gradients (HOG) to capture the distribution of the normal orientation of the surface in 4D space. Yang et al. [10] proposed to concatenate the normal vectors into a spatio-temporal sub-volume of depth together to capture more informative geometric clues. In the work of Lu et al. [11], to represent complex human activities involving human–object interactions without taking into account holistic human postures, discriminating local patterns were used, and also, the authors proposed to study the relationship between the sampled pixels in the actor and background regions. A common limitation of depth-based approaches is the view sensitivity and time consumption due to the heavy signature compared to skeleton-based approaches.

2.2. Skeleton-Based Representation

Human movements can be effectively captured by the positional dynamics of each skeletal joint [12–15], or the relationship between joint pairs [4,16], or even their combination [17–19]. In [8], a tool for monitoring the human skeleton (3D posture) in real time from an image at a single depth was developed. The existing skeleton-based human action recognition can be broadly grouped into two main categories: joint-based approaches and body part-based approaches. Joint-based approaches consider the human skeleton as a set of points, whereas body part-based approaches consider the human skeleton as a connected set of rigid segments between connected pairs of body parts. In [14], human skeletons were represented using the 3D joint locations, and a temporal hierarchy of co-variance descriptors was proposed to model joint trajectories. F.Lv and R.Nevatia in [15] proposed to use the hidden Markov models (HMMs) to represent the position of the joints. Devanne et al. [20] represented the 3D position evolution as a trajectory of movement. The problem of action recognition was then formulated as the problem of calculating the similarity between the shape of trajectories in a Riemannian manifold. Along similar lines, in these works [21,22] presented a Riemannian analysis of distance trajectories for real-time action recognition. In [23], the relative positions of pairwise articulations were used to represent the human skeleton, and the temporal evolution of this representation was modeled using a hierarchy of Fourier coefficients. X.Yang et al. in [16] proposed an effective method using the relative articular positions, temporal displacement of joints, and offset of the joints with respect to the initial frame.

The second category of skeleton-based approaches investigates the body parts. In [2], the human skeleton was represented by points in the Lie group $SE(3) \times \ldots \times SE(3)$, by explicitly modeling the 3D geometric relationships between various body parts within a frame using rotations and translations, then the human action was modeled as curves in this Lie group. The temporal evolution was handled by dynamic time warping (DTW). On the other hand, in [24], the human skeleton was hierarchically divided into smaller parts, and each part was represented using some bio-inspired features. Linear dynamic systems were used to model the temporal evolution of this part. Generally speaking, the joint-based method owns a faster calculation speed, while body the part-based method owns higher accuracy [25].

2.3. RGB-D-Based Development

The depth image is robust to lighting changes. However, it loses some useful information, such as texture context, which is essential to distinguish certain activities involving human-object interactions. Recently, several works showed that to improve the recognition of activities with object interactions, it is also necessary to merge RGB sequences with depth images [26–33]. For example, in the work of Zhao et al. [33], combined descriptors based on points of interest, extracted from RGB sequences and depth sequences, were brought together to perform the recognition. Liu and Shao [26] used a deep architecture to simultaneously merge RGB information and depth images; in [19], a set of random forests was used to merge spatio-temporal and human key articulations; Shahroudy et al. [27], used a structured density method by merging RGB information and skeleton indices, and in [29], the authors simply concatenated skeletal features and silhouette-based features to perform classification.

The review and analysis of current RGB-D action datasets revealed some limitations including size, applicability, availability of ground truth labels, and evaluation protocols. There is also the problem of dataset saturation, a phenomenon whereby the algorithms reported achieved a near-perfect performance.

3. Spatio-Temporal Modeling

3.1. Proposed Approach

In this work, we propose a framework for human activity recognition using the body part-based skeleton for action recognition and object detection and object tracking for human–object interaction recognition. Figure 1 summarizes the proposed approach: First, skeleton and object sequences are represented as trajectories in the Lie group, and these trajectories are then mapped into the Lie algebra, then to a Riemannian manifold to be compared in a rate-invariant way. In addition to distances to training trajectories, the output of the last layer of the neural network used for object detection is also used, in some scenarios, to build the final feature vector. The classification is therefore performed using the Hoeffding tree ("very fast decision trees (VFDT)").

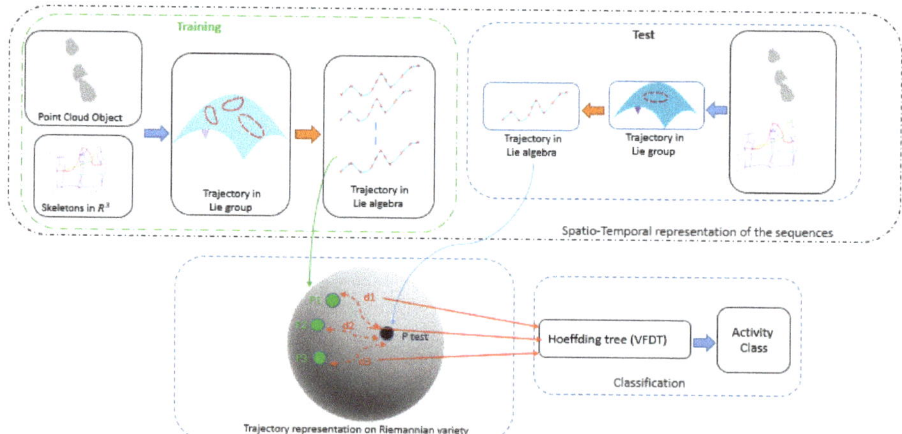

Figure 1. Overview of the proposed approach. VFDT, very fast decision trees.

Inspired by the work proposed in [2], which focused on the rigid transformations between different parts of the body within the same frame, we propose to model the evolution of the same part of the body (a segment) across frames by using the rotation and the translation necessary to transform the segment at frame t to the correspondent segment at frame $t+1$. This geometric representation of the rotation and translation of the rigid body in 3D space is part of the special euclidean group $SE(3)$ [34]. The evolution between two successive frames can be therefore modeled as a point in $SE(3) \times \ldots \times SE(3)$ $n-1$ times, where $n-1$ represents the number of body segments for a skeleton with n joints. A sequence of N frames is therefore represented by $N-1$ points in $SE(3) \times \ldots \times SE(3)$ ($n-1$ times) and can be modeled as a trajectory the in $SE(3) \times \ldots \times SE(3)$ ($n-1$ times) manifold. When an object is considered, an existing neural network is used for object detection in the first frame (RGB of the object in sequence i; frame j denoted by $OBJ - RGB(i)(j)$), then the object is tracked during the sequence (depth of the object in sequence i; frame j denoted by $OBJ - Depth(i)(j)$) using the iterative closest point (ICP) algorithm. The rigid deformations of the object across frames creates an additional trajectory in $SE(3)$ that is considered with the previous trajectory generated by the skeleton motion, to yield a final trajectory in $SE(3)\times\ldots\times SE(3)$ (n times). The next step is to map this trajectory to the corresponding Lie algebra $se(3)\times\ldots\times se(3)$, which is the tangent space at the identity element. The resulting trajectories lie in a euclidean space (Lie algebra) and incorporate the geometric deformations between body segments across frames. In order to compare their shapes independently of the execution rate, they are mapped to the shape space of continuous curves via the square root velocity manifold representation [35]. The classification is performed later using the Hoeffding tree (VFDT) [36] based on the elastic metric in the shape space.

3.2. Skeleton Motion Modeling

Firstly, we present the spatio-temporal modeling of the sequences. For this, we describe the geometric relation between the part of the body (denoted by *part*) at frame f_t and the same part in succession frame f_{t+1}. To do this, we use the rotation and translation required to move the current part to the position and orientation of the same part in the next frame, and we use the *procruste* function. This geometric transformation such as rotation and translation between two rigid body parts is a member of the special Euclidean group *SE(3)* [34] and defined by the following four by four matrix of the form:

$$P(R,\vec{d}) = \begin{bmatrix} R & \vec{d} \\ 0 & 1 \end{bmatrix} \in SE(3) \tag{1}$$

where $\vec{d} \in \mathbb{R}^3$ and $R \in \mathbb{R}^{3\times 3}$ is a rotation matrix, which is a point on the special orthogonal group $SO(3)$.

This geometrical transformation between two parts of the rigid body with two successive frames is represented by a point in $SE(3)$. Obviously, all parts of the body are presented by a point of the Lie group $SE(3)\times\ldots\times SE(3)$, where $\times$ denotes the direct product between Lie groups. Therefore, the temporal transformation of the body parts can be modeled by a trajectory in the $SE(3)\times\ldots\times SE(3)$ Lie group, as depicted in Figure 2.

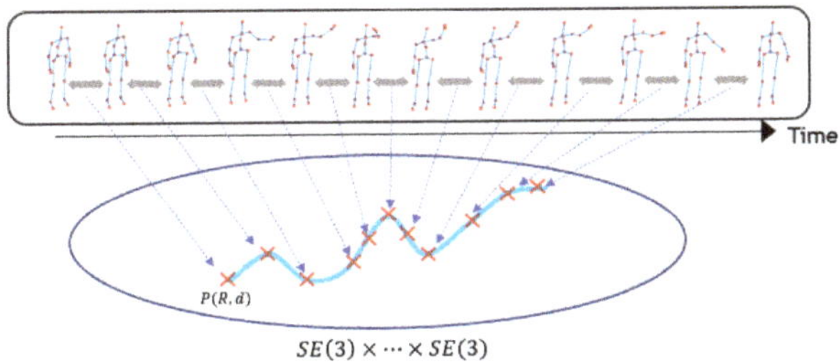

Figure 2. Action as a curve in the Lie group.

The Lie group identity element I_4 is defined by a four by four matrix. Mathematically, the tangent space to $SE(3)$ at the identity element is symbolized by $se(3)$, and it is considered to be the Lie algebra of $SE(3)$. This tangent space is a six-dimensional space constructed by matrices of the form:

$$B = \begin{bmatrix} U & \vec{w} \\ 0 & 0 \end{bmatrix} = \begin{bmatrix} 0 & -u_3 & u_2 & w_1 \\ u_3 & 0 & -u_1 & w_2 \\ -u_2 & u_1 & 0 & w_3 \\ 0 & 0 & 0 & 0 \end{bmatrix} \in se(3) \tag{2}$$

where $\vec{w} \in \mathbb{R}^3$ and $U \in \mathbb{R}^{3\times 3}$ the skew-symmetric matrix. Thus, it can be presented as a six-dimensional vector:

$$vec(B) = \begin{bmatrix} u_1, u_2, u_3, w_1, w_2, w_3 \end{bmatrix} \tag{3}$$

The exponential map for $SE(3)$ is defined as $\exp_{SE(3)} : se(3) \rightarrow SE(3)$ and the inverse exponential map, defined as $\log_{SE(3)} : SE(3) \rightarrow se(3)$. Both are used to navigate between the manifold and the tangent space, respectively, given by:

$$\exp_{SE(3)}(B) = e^B, \log_{SE(3)}(P) = log(P) \tag{4}$$

where e and log denote the matrix exponential and logarithm, respectively.

The geometric transformation between all the parts of two successive frames f_t and f_{t+1} can be represented as:

$\delta(t) = (P_{f_t(1),f_{t+1}(1)}(t), P_{f_t(2),f_{t+1}(2)}(t) \ldots, P_{f_t(M),f_{t+1}(M)}(t)) \in SE(3) \times \ldots \times SE(3)$, where M is the number of body parts. Using this representation, a skeletal sequence describes an action as a curve in $SE(3) \times \ldots \times SE(3)$. One can not directly classify the action curves in the curved space $SE(3) \times \ldots \times SE(3)$, according to [2]. In addition, temporal modeling approaches are not directly applicable to this space. For that, we will map the trajectory in $SE(3) \times \ldots \times SE(3)$ to its Lie algebra $se(3) \times \ldots \times se(3)$, the tangent space at the identity element I_4. With this method, we will map all the trajectories from the Lie group to the same tangent space to the identity, and we argue that the mapped curves are quite faithful to the original curves because they are close to the identity element of the Lie group as they represent the transformations of the same body parts across successive frames. The resulting curve in the Lie algebra corresponding to $\delta(t)$ is given by:

$$\begin{aligned} \sigma(t) = (&vec(log(P_{f_t(1),f_{t+1}(1)}(t))), vec(log(P_{f_t(2),f_{t+1}(2)}(t))) \\ &\ldots, vec(log(P_{f_t(M),f_{t+1}(M)}(t)))) \in se(3) \times \ldots \times se(3) \end{aligned} \tag{5}$$

The dimension of the characteristic vector at any time t of $\sigma(t)$ is equal to $6M$. For this, the temporal representation of the action sequence is a vector of dimension $6 \times M \times N$, where M is the number of parts ($M = 19$ parts), and N represents the number of frames in the sequence.

3.3. Object Modeling

3.3.1. Object Detection

In order to deal with human–object interactions, one key step is to recognize the object from an RGB image. For this, we propose to use an object recognition algorithm based on the neural network [37]. The you only look once (YOLO) model processes images in real time at 45 frames per second. A smaller version of the network, Fast YOLO, processes 155 frames per second while still achieving double the mAPof other real-time detectors. Compared to state-of-the-art detection systems, YOLO makes more localization errors, but is less likely to predict false positives on the background. YOLO system detection is a regression problem. It divides the image into an even grid S × S and simultaneously predicts bounding boxes B, confidence in those boxes, and class probabilities C. These predictions are encoded as an S × S × (B × 5 + C) tensor. YOLO imposes strong spatial constraints on the box prediction delineation since each cell in the grid predicts only two boxes and can only have one class. This spatial constraint limits the number of nearby objects that the model can predict. Figure 3 shows the steps of detecting objects in an RGB image.

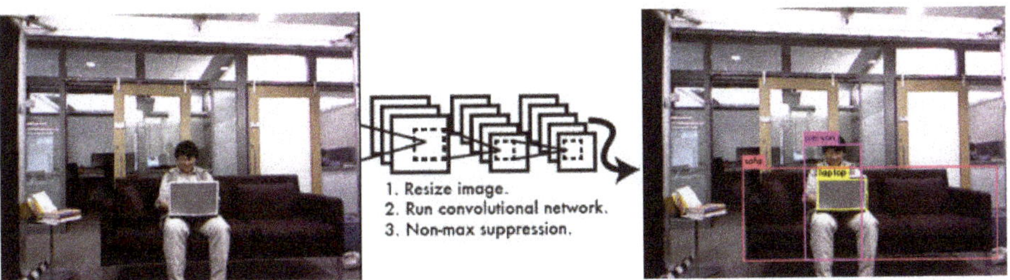

Figure 3. The YOLO detection system. (1) resizes the input image to 448 × 448; (2) runs a single convolutional network on the image; and (3) thresholds the resulting detections by the model's confidence.

3.3.2. Object Trajectory

Once having been detected in 2D, the object is tracked in 3D using the ICP algorithm. The resulting successive transformations are modeled as a trajectory in $SE(3)$. This trajectory is then mapped to the Lie algebra $se(3)$ and is fused with the trajectory in $se(3) \times \ldots \times se(3)$ ($n-1$ times) generated by the body parts. The trajectories modeling the activity lie in $se(3) \times \ldots \times se(3)$ (n times) and have to be compared independently of the execution rate. Therefore, they are considered as time parameterized curves, and an elastic metric will be used to provide a time re-parameterization-invariant metric. The additional trajectory (generated by the object) is used only when comparing the proposed approach to RGB-D-based approaches. In this case, the output of the last layer of the object detection neural network applied on the first frame is also used (when the color channel is considered) to build the final feature vector.

4. Rate Invariance Modeling and Classification

We start by outlining a mathematical framework for helping in analyzing the temporal evolution of human activity when viewed as trajectories on the shape space of parametrized curves. This framework respects the underlying geometry of the shape of the trajectories, seen as curves, and helps maintain the desired invariance, especially re-parameterization of the trajectory curve that represents the execution rate. The next step is to calculate the distance between a given trajectory (to classify) to all training ones; let k trajectories be in the training set, resulting in a k-dimensional feature vector.

4.1. Elastic Metric for Trajectories

This representation has been used previously in biometric and soft-biometric application [38–43]. In our case, we will analyze the shape of the trajectories by the square root velocity function (SRVF) $q : I \to \mathbb{R}^n$ defined as:

$$q(t) = \frac{\sigma(t)}{\sqrt{||\sigma(t)||}} \tag{6}$$

$q(t)$ is a special function introduced in [35] that captures the form of $\sigma(t)$ while offering easy calculations, and the L^2 norm represents the metric that allows us to compare the shape of two trajectories. The set of all trajectories, denoted as C, is thus defined as follows:

$$C = \{q : I \to \mathbb{R}^n | ||q|| = 1\} \subset \mathbb{L}^2(I, \mathbb{R}^n) \tag{7}$$

$||.||$ is the norm. With the norm on its tangent space, C becomes a Riemannian manifold named the pre-shape space. Each element represents a trajectory in $\mathbb{R}^n$. We define the distance between two elements q_1 and q_2 by the length of the geodesic path between q_1 and q_2 on the variety C. The geodesic path between any two points $q_1, q_2 \in C$ is given by the great circle, $\psi : [0, 1] \to C$, where:

$$\psi(\tau) = \frac{1}{\sin(\theta)} \times (\sin((1-\tau)\theta) \times q_1 + \sin(\theta\tau) \times q_2) \tag{8}$$

The geodesic length is $\theta = d_C(q_1, q_2) = cos^{-1}(< q_1, q_2 >)$. Let us define the equivalent class of q as: $[q] = \{\sqrt{\dot{\gamma}(t)} \times q(\gamma(t)),\ \ \gamma \in \Gamma\}$. The set of such equivalence classes, denoted by $\mathcal{S} \doteq \{[q] | q \in \mathcal{C}\}$, is called the shape space of open curves in $\mathbb{R}^n$. As shown in [35], $\mathcal{S}$ inherits a Riemannian metric from the larger space $\mathcal{C}$ due to the quotient structure. To obtain geodesics and geodesic distances between elements of $\mathcal{S}$, one needs to solve the optimization problem:

$$\gamma^* = argmin_{\gamma \in \Gamma} d_c(q_1, \sqrt{\dot{\gamma}} \times (q_2 \circ \gamma)). \tag{9}$$

The optimization over Γ is done using the dynamic programming algorithm. Let $q_2^*(t) = \sqrt{\dot{\gamma^*(t)}} \times q_2(\gamma^*(t)))$ be the optimal element of $[q_2]$, associated with the optimal re-parameterization γ^* of the second curve, then the geodesic distance between $[q_1]$ and $[q_2]$ in $\mathcal{S}$ is $d_s([q_1], [q_2]) \doteq d_c(q_1, q_2^*)$, and the geodesic is given by Equation (8), with q_2 replaced by q_2^*.

4.2. Feature Vector Building and Classification

We propose four variants of our method based on the channels used. The first one (geometric G) uses only the skeleton data. The second one (G + D) uses the skeleton and the depth channel. The third variant (G + C) uses the skeleton and the color channels. The last variant uses all channels (G + D + C). We present first the feature vector for the geometric G approach. Let n be the number of joints in a skeleton, the spatio-temporal modeling presented in Section 3, and k the number of trajectories in the training set with labels $l_1, \ldots, l_k$. For a given sequence in the test set, the first step is to represent it as a trajectory in $SE(3)^{n-1}$ as described in Section 3. Then, the elastic framework is applied in order to compute the elastic distance from the given trajectory to each of the k training ones. As illustrated in the previous section, the use of the elastic metric in trajectories' comparison ensures a rate-invariant distance. The resulting vector of k distances represents the feature vector of the geometric approach (G). An additional trajectory in $SE(3)$ must be considered when the depth data are used in the (G + D) approach. The feature vector has the same size; however, the trajectories are considered in $SE(3)^n$ rather than $SE(3)^{n-1}$. When the color channel is considered, the output of the last layer in the deep network used for object detection is concatenated to the k distance in order to build the feature vector denoted by $FeatureV$. The steps of feature vector building and classification are illustrated in Algorithm 1.

The resulting feature vector is fed to the Hoeffding tree (VFDT) algorithm. The Hoeffding tree [36] or very fast decision tree (VFDT) is built incrementally over time by splitting nodes (into two) using a small amount of the incoming data stream. The number of samples considered by the learning to expand a node depends on a statistical method called the Hoeffding bound or additive Chernoff bound. The Hoeffding tree is constructed by making recursive splits of leaves from a blank root and subsequently getting internal decision nodes, such that a tree structure is formed. The splits are decided by heuristic evaluation functions that evaluate the merit of the split-test based on attribute values.

Algorithm 1 Action sequences' classification.

Input:
k+1 sequences: $S_1, \ldots, S_k$ k training sequences of size $w_1, \ldots, w_k$, respectively, and S_{k+1} test sequence of size w_{k+1}.
r: number of body parts (r = n − 1).
Output:
Label of sequence S_{k+1}
Begin
for $i = 1$ to $k+1$ **do**
  OBJ-RGB(i)(1) = YOLO(S_i(1))
  $OBJ - Depth(i)(1) = RGB2Depth(OBJ - RGB(i)(1))$
  for $j = 1$ to $w_i - 1$ **do**
    $(R_{i,j}, d_{i,j})$ = ICP (OBJ-Depth(i)(j), OBJ-Depth(i)(j+1))
    for $l = 1$ to r **do**
      $P_{i,j,l} = (R_{i,j,l}, d_{i,j,l}) = procruste(part_{i,j,l}, part_{i,j+1,l})$
      $\sigma(i,j,l) = vec(log_{SE(3)}(P_{i,j,l}))$
    end for
  end for
end for
$q_{k+1} = \frac{\sigma(k+1)}{\sqrt{\|(\sigma(k+1,))\|}}$
for $i = 1$ to k **do**
  $q_i = \frac{\sigma(i)}{\sqrt{\|(\sigma(i))\|}}$
  $\gamma^* = argmin_{\gamma \in \Gamma} d_c(q_1, \sqrt{\dot{\gamma}}(q_2 \circ \gamma))$.
  $q_i^* = \sqrt{\dot{\gamma}^*}.q_i(\gamma^*))$
  $d_i = d(S_i, S_k + 1) = cos^{-1}(< q_i^*, q_{k+1} >)$
end for
FeatureV = concatenate($d_1, .., d_k, OBJ - RGB(k+1)(1)$)
label ($S_k + 1$) = VFDT (FeatureV)
End

5. Experimentation and Results

In order to validate our method, an evaluation was conducted on three databases that represent different challenges, namely Microsoft Research (MSR) Action3D dataset [8], MSR-Daily Activity 3D [23], and the SYSU 3D Human-Object Interaction Set [44].

5.1. MSR Action 3D

5.1.1. Data Description and Protocol

The MSR-Action 3D dataset is a set of RGBD data captured by a Kinect. This dataset includes 20 actions performed by 10 different subjects facing the camera. Each action is performed two or three times, resulting in a total of 557 action sequences. 3D joint positions are extracted from the depth sequence using the real-time skeleton tracking algorithm proposed in [45]. All actions are performed without interaction with the objects. Two main challenges are identified: the strong similarity between the different groups of actions and the changes in the speed of execution of the actions. For each sequence, the dataset provides information about depth, color, and skeleton. As indicated in [8], ten sequences are not used in the experiments because the skeletons are missing or too erroneous. For our experiments, we use 547 sequences. In this dataset, we followed the same protocol of the cross topic of [8], in which half of the subjects are used for training and the other half for testing. Subjects 1, 3, 5, 7, and 9 are used for training and Subjects 2, 4, 6, 8, and 10 for testing. In [8], the sequences were divided into three subsets AS_1, AS_2, and AS_3, each containing eight actions. Sets AS_1 and AS_2 are intended to group actions with similar movements, while AS_3 is intended to group complex actions.

5.1.2. Experimental Result and Comparison

No action, in this dataset, includes an interaction with the object. Thus, the skeleton-based approach (G) is performed. Table 1 reports the recognition performance on MSR-Action3D compared to several state-of-the-art approaches: joint positions (JPs) [14]: concatenation of the 3D coordinates of all the joints $v_1, \ldots, v_N$; pairwise relative positions of the joints (RJPs) [16]: concatenation of all the vectors; joint angles (JAs) [5]: concatenation of the quaternions corresponding to all joint angles (we also tried Euler angles and Euler axis-angle representations for the joint angles, but quaternions gave the best results); individual body part locations (BPLs) [46]: each individual body part is represented as a point in $SE(3)$ using its rotation and translation relative to the global x-axis.

In the last row of Table 1, the average recognition rate for the three subsets AS_1, AS_2, and AS_3 is reported. The recognition rates of our approach on AS_1, AS_2, and AS_3 were 94.66%, 85.08%, and 96.76%, respectively. The accuracy on subset AS_2 was lower than the two other subsets. This behavior is similar to the state-of-the-art approaches as revealed in Table 1. The average accuracy of the proposed representations was 92.16%, which is superior to the performance of previous state-of-the-art approaches provided in Table 1.

Table 2 compares the proposed approach with various approaches to recognizing human actions on skeletons using the protocol of [8]. Here, we see that our approach is competitive with the state-of-the-art with a recognition rate equal to 92.16%.

Table 1. Recognition performance on the MSR-Action3D for different feature spaces using the protocol of [8]. JP, joint position; RJP, relative position of the joint; JA, joint angle; BPL, body part location.

Dataset	JP [14]	RJP [16]	JA [5]	BPL [46]	Proposed
AS_1	91.65	92.15	85.80	83.87	**94.66**
AS_2	75.36	79.24	65.47	75.23	**85.08**
AS_3	94.64	93.31	94.22	91.54	**96.76**
Average	87.22	88.23	81.83	83.54	**92.16**

Table 2. Comparison with the state-of-the-art results.

MSR-Action3D Dataset (Protocol of [8])	
Histograms of 3D joints [47]	78.97
EigenJoints [16]	82.30
Joint angle similarities [5]	83.53
Spatial and temporal part sets [48]	90.22
Co-variance descriptors [14]	90.53
Random forests [19]	90.90
Body parts (BP)+SRVF [20]	92.10
Intra-frame modeling [2]	92.49
Proposed approach: skeleton	**92.16**

5.2. MSR Daily Activity 3D

5.2.1. Data Description and Protocol

The MSR Daily Activity 3D dataset [23] is a set of RGB-D sequences of human sequences acquired with the Kinect. It contains 16 types of activities: drink, eat, read book, call cellphone, write on a paper, use laptop, use vacuum cleaner, cheer up, sit still, toss paper, play game, lay down on sofa, walk, play guitar, stand up, sit down. Each of them was performed twice by 10 subjects [23]. The dataset contains 320 videos = 16 × 10 × 2 (10 actors and two essays/actor). There are 20 body joints recorded, whose positions are quite noisy due to two poses: "sitting on sofa" and "standing close to sofa".

The experimental protocol is the same as in [23], which divides the dataset into three subsets, AS_1, AS_2, and AS_3, as shown in Table 3.

Table 3. Subsets of actions, AS_1, AS_2, and AS_3 in the MSR Daily Activity 3D dataset [23].

AS_1	AS_2	AS_3
eat	drink	use laptop
read book	call cellphone	cheer up
write on a paper	use vacuum cleaner	play guitar
use laptop	sit still	stand up
toss paper	play game	sit down
walk	lie down on sofa	

5.2.2. Experimental Result and Comparison

Table 4 reports the results of our algorithm on the MSR Daily activity dataset. The average recognition rate using only the skeleton data is 87.55%. When the dynamics of the object is considered, we have an average recognition rate equal to 88%. Combining the feature vector resulting from the geometry of the skeleton and object to the appearance of the object yields good improvement of the recognition rate. Actually, using the geometry of the skeleton (G) and the appearance of the object (C), the average recognition rate is 94.44%. The performance is also improved by using in addition the geometry of the object (D) to reach a 95% recognition rate, which is very competitive compared with recent state-of-the-art approaches.

Table 4. Recognition performance on the MSR-DailyActivity3D dataset for different feature spaces: (D) depth; (C) color (or RGB); (G) geometry or skeleton.

Methods	**Accuracy %**
(G) Dynamic Temporal Warping [49]	54
(G) 3D Joints and Local occupancy patterns (LOP) [50]	78
(G) Histogram of Oriented 4D Normals (HON4D) [3]	80.00
(G) Spar-Sity learning to Fuse atomic Features (SSFF) [27]	81.9
(G) Deep Model-Restricted Graph-based Genetic Programming (RGGP) [26]	85.6
(G) Action-let Ensemble [50]	85.75
(G) Super Normal [10]	86.25
(G) Bilinear [51]	86.88
(G) Depth Cuboid Similarity Feature (DCSF) + Joint [52]	88.2
(G) Local Flux Feature (LFF) + Improved Fisher Vector (IFV) [28]	91.1
(G) Group Sparsity [12]	95
(G) Range Sample [11]	95.6
(G) Heterogeneous Feature Machines (HFM) [53]	84.38
(G) Model of Probabilistic Canonical Correlation Analyzers (MPCCA) [54]	90.62
(G) Multi-Task Discriminant Analysi (MTDA) [55]	90.62
(G + D + C) JOULE [44]	95
Our Method:(G) Skeleton	**87.55**
Our Method:(G + D) Skeleton + Obj(D)	**88**
Our Method:(G + C) Skeleton + Obj(RGB)	**94.44**
Our Method:(G + D + C) Skeleton + Obj(RGB) + Obj(D)	**95**

5.3. *SYSU 3D Human-Object Interaction Set*

5.3.1. Data Description and Protocol

In this dataset [44], twelve different activities focusing on interactions with objects were performed by 40 persons. For each activity, each participant manipulates one of the six different objects: phone, chair, bag, wallet, mop, and besom. Therefore, there are in total 480 video clips collected in this set. For each video clip, the data acquisition is done by a Kinect camera, and we have the corresponding

RGB images, the depth sequence, and the skeleton. We tested all the methods compared with the second setting (Setting 2) [44]. The video footage made by half of the participants was used to learn the parameter model and the rest for the tests. We report the average precision and the standard deviation of the results on 30 random divisions For each parameter.

5.3.2. Experimental Result and Comparison

Table 5 provides the results of the different variants of the proposed approach compared to the state-of-the-art. When only the geometry of the skeleton data is considered, the recognition rate is 73.48%. This result is competitive compared to previous geometric approaches (based only on skeleton data). If we take into account the dynamics of the object, the recognition rate is equal to 74.51%. When the appearance of the object is considered in addition to the geometry of the skeleton, the recognition rate reaches 86.76 %, which represents the best recognition rate compared to the recent state-of-the-art approaches. The full version of the proposed approach, which makes use of all RGB-D and skeleton information, provides a recognition rate of 87.40%.

For further analysis of the obtained results, we illustrate in Figures 4 and 5 the confusion matrices on the SYSU dataset using the skeleton data (G) and the RGB-D (G + D + C) channels, respectively. The appearance of the object improves the performance of all actions, but the improvement is more considerable for sweeping and mopping actions: the skeleton data performs good for several actions, but seems not sufficient to distinguish other actions such as sweeping, with recognition rates of 35% and 54.9%, respectively. The skeleton motion during these two actions is similar to the motion while drinking, moving the chair, or pouring. The appearance of the object improves the performance for the sweeping and mopping actions. As shown in Figure 5, the recognition of sweeping improved by 36% by using the geometry and the appearance of the object in addition to the skeleton motion. The performance of mopping reaches 78.3% for the mopping action.

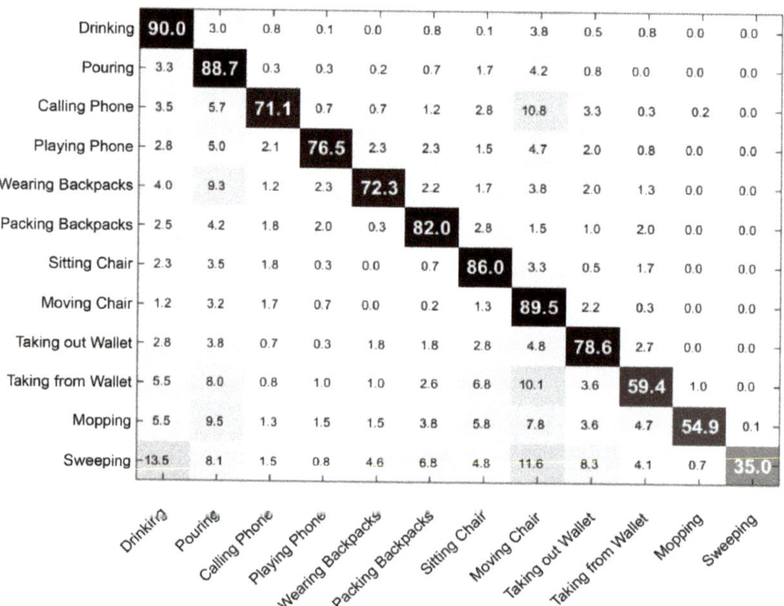

Figure 4. SYSU 3D Human-Object Interaction dataset confusion matrix based on skeleton data (G).

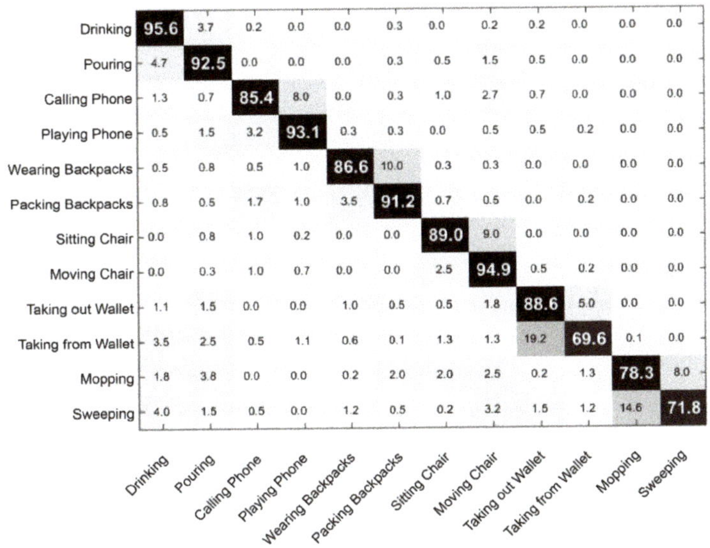

Figure 5. SYSU 3D Human-Object Interaction dataset confusion matrix based on skeleton, depth, and color data (G + D + C).

Table 5. Comparison on the SYSU 3D dataset. (D) depth; (C) color (or RGB); (G) geometry or skeleton.

Methods	Accuracy %
(G) Local Accumulative Frame Feature (LAFF) [56]	54.2
(G) Dynamic skeletons [44]	75.5 ± 3.08
(G) LSTM-trust gate [57]	76.5
(G + D + C) LAFF [56]	80
(G + D + C) JOULE [44]	84.9 ± 2.29
Our Method:(G) Skeleton	**73.48 ± 5.91**
Our Method:(G + D) Skeleton + Obj (D)	**74.51 ± 5.47**
Our Method:(G + C) Skeleton + Obj (RGB)	**86.76 ± 4.82**
Our Method:(G + D + C) Skeleton + Obj (RGB) + Obj (D)	**87.40 ± 5.04**

6. Conclusions and Future Direction

In this paper, we represent the inter-frame evolution of skeleton body parts in Lie group $SE(3) \times \ldots \times SE(3)$. When an object is involved in the action, a neural network is used to detect the object at the first frame, then the evolution across frames is tracked, then similarly modeled as an additional trajectory in Lie group $SE(3)$. The resulting trajectories are then mapped onto the Lie algebra, where they are compared using a re-parameterization-invariant framework in order to handle rate variations. The distances to training trajectories are concatenated with the output of the last layer of the neural network used for object detection, then are fed to the very fast decision tree to perform action recognition. We experimentally show that the proposed approach performs better than many previous approaches for human activity recognition. As future work, we expect widespread applicability in domains such as physical therapy and rehabilitation.

Author Contributions: Formal analysis, H.D. and I.R.F.; Methodology, M.B., H.D. and I.R.F.; Supervision, M.M. and I.R.F.; Validation, H.D., M.M. and I.R.F.; Visualization, I.R.F.; Writing–original draft, M.B. All authors have read and agreed to the published version of the manuscript.

Funding: This research received no external funding.

Conflicts of Interest: The authors declare no conflict of interest.

References

1. Shotton, J.; Sharp, T.; Kipman, A.; Fitzgibbon, A.; Finocchio, M.; Blake, A.; Cook, M.; Moore, R. Real-time human pose recognition in parts from single depth images. *Commun. ACM* **2013**, *56*, 116–124. [CrossRef]
2. Vemulapalli, R.; Arrate, F.; Chellappa, R. Human Action Recognition by Representing 3D Skeletons as Points in a Lie Group. In Proceedings of the IEEE Conference on Computer Vision and Pattern Recognition (CVPR), Columbus, OH, USA, 23–28 June 2014.
3. Oreifej, O.; Liu, Z. Hon4d: Histogram of oriented 4d normals for activity recognition from depth sequences. In Proceedings of the IEEE International Conference on Computer Vision and Pattern Recognition, Portland, OR, USA, 23–28 June 2013; pp. 716–723.
4. Ofli, F.; Chaudhry, R.; Kurillo, G.; Vidal, R.; Bajcsy, R. Sequence of the most informative joints (smij): A new representation for human skeletal action recognition. *J. Vis. Commun. Image Represent.* **2014**, *25*, 24–38. [CrossRef]
5. Ohn-bar, E.; Trivedi, M.M. Joint Angles Similarities and HOG for Action Recognition. In Proceedings of the 2013 IEEE Conference on Computer Vision and Pattern Recognition Workshops, Portland, OR, USA, 23–28 June 2013.
6. Vemulapalli, R.; Chellapa, R. Rolling Rotations for Recognizing Human Actions From 3D Skeletal Data. In Proceedings of the IEEE Conference on Computer Vision and Pattern Recognition (CVPR), Las Vegas, NV, USA, 27–30 June 2016; pp. 4471–4479.
7. Boujebli, M.; Drira, H.; Mestiri, M.; Farah, I.R. Rate invariant action recognition in Lie algebra. In Proceedings of the 2017 International Conference on Advanced Technologies for Signal and Image Processing (ATSIP), Fez, Morocco, 22–24 May 2017.
8. Li, W.; Zhang, Z.; Liu, Z. Action recognition based on a bag of 3d points. In Proceedings of the IEEE International Conference on Computer Vision and Pattern Recognition Workshops, San Francisco, CA, USA, 13–18 June 2010; pp. 9–14.
9. Zhu, Y.; Chen, W.; Guo, G. Evaluating spatiotemporal interest point features for depth-based action recognition. *Image Vis. Comput.* **2014**, *32*, 453–464. [CrossRef]
10. Yang, X.; Tian, Y. Super normal vector for activity recognition using depth sequences. In Proceedings of the IEEE International Conference on Computer Vision and Pattern Recognition, Columbus, OH, USA, 24–27 June 2014; pp. 804–811.
11. Lu, C.; Jia, J.; Tang, C.-K. Range-sample depth feature for action recognition. In Proceedings of the IEEE International Conference on Computer Vision and Pattern Recognition, Columbus, OH, USA, 24–27 June 2014; pp. 772–779.
12. Luo, J.; Wang, W.; Qi, H. Group sparsity and geometry constrained dictionary learning for action recognition from depth maps. In Proceedings of the 2013 IEEE International Conference on Computer Vision, Sydney, Australia, 1–8 December 2013.
13. Du, Y.; Wang, W.; Wang, L. Hierarchical recurrent neural network for skeleton based action recognition. In Proceedings of the IEEE International Conference on Computer Vision and Pattern Recognition, Boston, MA, USA, 7–12 June 2015; pp. 1110–1118.
14. Hussein, M.E.; Torki, M.; Gowayyed, M.A.; El-Saban, M. Human action recognition using a temporal hierarchy of covariance descriptors on 3d joint locations. In Proceedings of the International Joint Conference on Artificial Intelligence, Beijing, China, 3–9 August 2013; pp. 2466–2472.
15. Lv, F.; Nevatia, R. Recognition and segmentation of 3-d human action using hmm and multi-class adaboost. In Proceedings of the European Conference on Computer Vision, Graz, Austria, 7–13 May 2006; pp. 359–372.
16. Yang, X.; Tian, Y. Eigenjoints-based action recognition using naivebayes-nearest-neighbor. In Proceedings of the IEEE International Conference on Computer Vision and Pattern Recognition Workshops, Providence, RI, USA, 16–21 June 2012; pp. 14–19.
17. Lillo, I.; Soto, A.; Niebles, J.C. Discriminative hierarchical modeling of spatio-temporally composable human activities. In Proceedings of the IEEE International Conference on Computer Vision and Pattern Recognition, Columbus, OH, USA, 23–28 June 2014; pp. 812–819.
18. Zanfir, M.; Leordeanu, M.; Sminchisescu, C. The moving pose: An efficient 3d kinematics descriptor for low-latency action recognition and detection. In Proceedings of the IEEE International Conference on Computer Vision, Sydney, Australia, 1–8 December 2013; pp. 2752–2759.

19. Zhu, Y.; Chen, W.; Guo, G. Fusing spatio-temporal features and joints for 3d action recognition. In Proceedings of the IEEE International Conference on Computer Vision and Pattern Recognition Workshops, Portland, OR, USA, 23–28 June 2013; pp. 486–491.
20. Devanne, M.; Wannous, H.; Berretti, S.; Pala, P.; Daoudi, M.; Del Bimbo, A. 3D Human Action Recognition by Shape Analysis of Motion Trajectories on Riemannian Manifold. *IEEE Trans. Cybern.* **2015**, *45*, 1340–1352. [CrossRef] [PubMed]
21. Meng, M.; Drira, H.; Boonaert, J. Distances evolution analysis for online and off-line human–object interaction recognition. *Image Vision Comput.* **2018**, *70*, 32–45. [CrossRef]
22. Meng, M.; Drira, H.; Daoudi, M.; Boonaert, J. Human-object interaction recognition by learning the distances between the object and the skeleton joints. In Proceedings of the 2015 11th IEEE International Conference and Workshops on Automatic Face and Gesture Recognition (FG), Ljubljana, Slovenia, 4–8 May 2015.
23. Wang, J.; Liu, Z.; Wu, Y.; Yuan, J. Mining Actionlet Ensemble for Action Recognition with Depth Cameras. In Proceedings of the IEEE International Conference, Providence, RI, USA, 16–21 June 2012.
24. Chaudhry, R.; Ofli, F.; Kurillo, G.; Bajcsy, R.; Vidal, R. Bio-inspired Dynamic 3D Discriminative Skeletal Features for Human Action Recognition. In Proceedings of the 2013 IEEE Conference on Computer Vision and Pattern Recognition Workshops, Portland, OR, USA, 23–28 June 2013.
25. Guo, S.; Pan, H.; Tan, G.; Chen, L.; Gao, C. A High Invariance Motion Representation for Skeleton-Based Action Recognition. *Int. J. Pattern Recognit. Artif. Intell.* **2016**, *30*, 1650018. [CrossRef]
26. Liu, L.; Shao, L. Learning discriminative representations from rgb-d video data. In Proceedings of the International Joint Conference on Artificial Intelligence, Beijing, China, 3–9 August 2013; pp. 1493–1500.
27. Shahroudy, A.; Wang, G.; Ng, T.-T. Multi-modal feature fusion for action recognition in rgb-d sequences. In Proceedings of the International Symposium on Control, Communications, and Signal Processing, Athens, Greece, 21–23 May 2014; pp. 1–4.
28. Yu, M.; Liu, L.; Shao, L. Structure-preserving binary representations for rgb-d action recognition. *IEEE Trans. Pattern Anal. Mach. Intell.* **2015**, *38*, 1651–1664. [CrossRef] [PubMed]
29. Chaaraoui, A.A.; Padilla-Lopez, J.R.; Florez-Revuelta, F. Fusion of skeletal and silhouette-based features for human action recognition with rgb-d devices. In Proceedings of the IEEE International Conference on Computer Vision Workshops, Sydney, Australia, 1–8 December 2013; pp. 91–97.
30. Koppula, H.S.; Gupta, R.; Saxena, A. Learning human activities and object affordances from rgb-d videos. *Int. J. Robot. Res.* **2013**, *32*, 951–970. [CrossRef]
31. Lei, J.; Ren, X.; Fox, D. Fine-grained kitchen activity recognition using rgb-d. In Proceedings of the ACM Conference on Ubiquitous Computing, Pittsburgh, PA, USA, 5–8 September 2012; pp. 208–211.
32. Wen, Z.; Yin, W. A feasible method for optimization with orthogonality constraints. *Math. Program.* **2013**, *142*, 397–434. [CrossRef]
33. Zhao, Y.; Liu, Z.; Yang, L.; Cheng, H. Combing rgb and depth map features for human activity recognition. In Proceedings of the IEEE Asia-Pacific Signal & Information Processing Association Annual Summit and Conference (APSIPA ASC), Hollywood, CA, USA, 3–6 December 2012; pp. 1–4.
34. Murray, R.M.; Li, Z.; Sastry, S.S. *A Mathematical Introduction to Robotic Manipulation*; CRC Press: Boca Raton, FL, USA, 1994.
35. Joshi, S.H.; Klassen, E.; Srivastava, A.; Jermyn, I. A novel representation for riemannian analysis of elastic curves in Rn. In Proceedings of the IEEE International Conference on Computer Vision and Pattern Recognition, Minneapolis, MN, USA, 17–22 June 2007; pp. 1–7.
36. Xu, W.; Qin, Z. Constructing Decision Trees for Mining High-speed Data Streams. *Chin. J. Electron.* **2012**, *21*, 215–220.
37. Redmon, J.; Divvala, S.; Girshick, R.; Farhadi, A. You Only Look Once: Unified, Real-Time Object Detection. In Proceedings of the IEEE Conference on Computer Vision and Pattern Recognition (CVPR), Las Vegas, NV, USA, 27–30 June 2016; pp. 779–788.
38. Drira, H.; Ben Amor, B.; Srivastava, A.; Daoudi, M.; Slama, R. 3D Face Recognition under Expressions, Occlusions, and Pose Variations. *IEEE Trans. Pattern Anal. Mach. Intell.* **2013**, *35*, 2270–2283. [CrossRef] [PubMed]
39. Xia, B.; Ben Amor, B.; Drira, H.; Daoudi, M.; Ballihi, L. Combining face averageness and symmetry for 3D-based gender classification. *Pattern Recognit.* **2015**, *48*, 746–758. [CrossRef]

40. Xia, B.; Amor, B.B.; Drira, H.; Daoudi, M.; Ballihi, L. Gender and 3D facial symmetry: What's the relationship? In Proceedings of the 2013 10th IEEE International Conference and Workshops on Automatic Face and Gesture Recognition (FG), Shanghai, China, 22–26 April 2013.
41. Amor, B.B.; Drira, H.; Ballihi, L.; Srivastava, A.; Daoudi, M. An experimental illustration of 3D facial shape analysis under facial expressions. *Ann. Telecommun.* **2009**, *64*, 369–379 [CrossRef]
42. Mokni, R.; Drira, H.; Kherallah, M. Combining shape analysis and texture pattern for palmprint identification. *Multimed. Tools Appl.* **2017**, *76*, 23981–24008 [CrossRef]
43. Xia, B.; Amor, B.B.; Daoudi, M.; Drira, H. Can 3D Shape of the Face Reveal your Age? In Proceedings of the 2014 International Conference on Computer Vision Theory and Applications (VISAPP), Lisbon, Portugal, 5–8 January 2014; pp. 5–13.
44. Hu, J.-F.; Zheng, W.-S.; Lai, J.; Zhang, J. Jointly learning heterogeneous features for rgb-d activity recognition. *IEEE Trans. Pattern Anal. Mach. Intell.* **2017**, *39*, 2186–2200. [CrossRef] [PubMed]
45. Shotton, J.; Fitzgibbon, A.; Cook, M.; Sharp,T.; Finocchio, M.; Moore, R.; Kipman, A.; Blake, A. Real-time human pose recognition in parts from single depth images. In Proceedings of the CVPR, Providence, RI, USA, 20–25 June 2011.
46. Yacoob, Y.; Black, M.J. Parameterized Modeling and Recognition of Activites. In Proceedings of the Sixth International Conference on Computer Vision, Bombay, India, 4–7 January 1998.
47. Xia, L.; Chen, C.C.; Aggarwal, J.K. View Invariant Human Action Recognition Using Histograms of 3D Joints. In Proceedings of the 2012 IEEE Computer Society Conference on Computer Vision and Pattern Recognition Workshops, Providence, RI, USA, 16–21 June 2012.
48. Wang, C.; Wang, Y.; Yuille, A.L. An Approach to Pose-based Action Recognition. In Proceedings of the 2013 IEEE Conference on Computer Vision and Pattern Recognition, Portland, OR, USA, 23–28 June 2013.
49. Muller, M.; Roder, T. Motion templates for automatic classification and retrieval of motion capture data. In Proceedings of the ACM SIGGRAPH/Eurographics Symposium on Computer Animation, Vienna, Austria, 2–4 September 2006; pp. 137–146.
50. Wang, J.; Liu, Z.; Wu, Y.; Yuan, J. Learning actionlet ensemble for 3d human action recognition. *IEEE Trans. Pattern Anal. Mach. Intell.* **2014**, *36*, 914–927. [CrossRef] [PubMed]
51. Kong, Y.; Fu, Y. Bilinear heterogeneous information machine for rgbd action recognition. In Proceedings of the IEEE International Conference on Computer Vision and Pattern Recognition, Boston, MA, USA, 7–12 June 2015; pp. 1054–1062.
52. Xia, L.; Aggarwal, J. Spatio-temporal depth cuboid similarity feature for activity recognition using depth camera. In Proceedings of the IEEE International Conference on Computer Vision and Pattern Recognition, Portland, OR, USA, 23–28 June 2013; pp. 2834–2841.
53. Cao, L.; Luo, J.; Liang, F.; Huang, T.S. Heterogeneous feature machines for visual recognition. In Proceedings of the IEEE International Conference on Computer Vision, Kyoto, Japan, 29 September–2 October 2009; pp. 1095–1102.
54. Cai, Z.; Wang, L.; Qiao, X.P.Y. Multi-view super vector for action recognition. In Proceedings of the IEEE Conference on Computer Vision and Pattern Recognition, Columbus, OH, USA, 24–27 June 2014; pp. 596–603.
55. Zhang, Y.; Yeung, D.-Y. Multi-task learning in heterogeneous feature spaces. In Proceedings of the Conference on Artificial Intelligence, San Francisco, CA, USA, 7–11 August 2011.
56. Hu, J.-F.; Zheng, W.-S.; Ma, L.; Wang, G.; Lai, J. Real-time RGB-D activity prediction by soft regression. In Proceedings of the European Conference on Computer Vision (ECCV), Amsterdam, The Netherlands, 8–16 October 2016; pp. 280–296.
57. Liu, J.; Shahroudy, A.; Xu, D.; Wang, G. *Spatio-Temporal LSTM with Trust Gates for 3D Human Action Recognition*; Springer: Cham, Switzerland, 2016; pp. 816–833.

Publisher's Note: MDPI stays neutral with regard to jurisdictional claims in published maps and institutional affiliations.

Article

A Statistical Approach for the Assessment of Muscle Activation Patterns during Gait in Parkinson's Disease

Giulia Pacini Panebianco [1,2,*], Davide Ferrazzoli [3], Giuseppe Frazzitta [4], Margherita Fonsato [5], Maria Cristina Bisi [1], Silvia Fantozzi [1,2] and Rita Stagni [1,2]

1 Department of Electric, Electronic and Information Engineering "Guglielmo Marconi"—DEI, University of Bologna, Viale Risorgiento 2, 40136 Bologna, Italy; mariacristina.bisi@unibo.it (M.C.B.); silvia.fantozzi@unibo.it (S.F.); rita.stagni@unibo.it (R.S.)

2 CIRI Health Sciences & Technologies, Via Tolara di Sopra, 50, 40064 Ozzano dell'Emilia, Italy

3 Department of Neurorehabilitation, Hospital of Vipiteno (SABES-ASDAA), Via Santa Margherita, 24, 39049 Vipiteno-Sterzing, Italy; davideferrazzoli@gmail.com

4 MIRT Park Project, Vicolo Carnaro, 10, 91025 Marsala, Italy; frazzittag62@gmail.com

5 "Moriggia-Pelascini" Hospital, Via Pelascini, 3, 22015 Gravedona ed Uniti, Italy; margherita.fonsato@gmail.com

* Correspondence: giulia.pacini2@unibo.it

Received: 31 August 2020; Accepted: 1 October 2020; Published: 5 October 2020

Abstract: Recently, the statistical analysis of muscle activation patterns highlighted that not only one, but several activation patterns can be identified in the gait of healthy adults, with different occurrence. Although its potential, the application of this approach in pathological populations is still limited and specific implementation issues need to be addressed. This study aims at applying a statistical approach to analyze muscle activation patterns of gait in Parkinson's Disease, integrating gait symmetry and co-activation. Surface electromyographic signal of tibialis anterior and gastrocnemius medialis were recorded during a 6-min walking test in 20 patients. Symmetry between right and left stride time series was verified, different activation patterns identified, and their occurrence (number and timing) quantified, as well as the co-activation of antagonist muscles. Gastrocnemius medialis presented five activation patterns (mean occurrence ranging from 2% to 43%) showing, with respect to healthy adults, the presence of a first shorted and delayed activation (between flat foot contact and push off, and in the final swing) and highlighting a new second region of anticipated activation (during early/mid swing). Tibialis anterior presented five activation patterns (mean occurrence ranging from 3% to 40%) highlighting absent or delayed activity at the beginning of the gait cycle, and generally shorter and anticipated activations during the swing phase with respect to healthy adults. Three regions of co-contraction were identified: from heel strike to mid-stance, from the pre- to initial swing, and during late swing. This study provided a novel insight in the analysis of muscle activation patterns in Parkinson's Disease patients with respect to the literature, where unique, at times conflicting, average patterns were reported. The proposed integrated methodology is meant to be generalized for the analysis of muscle activation patterns in pathologic subjects.

Keywords: surface EMG; statistical gait analysis; activation patterns; co-activation; Parkinson's disease

---

1. Introduction

Surface electromyography (sEMG) is widely used in clinical gait analysis [1,2]: amplitude-, time-, frequency-based parameters, and conduction velocity of muscle fibers can be investigated to characterize muscle activity from sEMG [3,4]. In particular, temporal parameters, i.e., the on/off timing of muscle activation, have become a de-facto standard for the clinical assessment of gait [5–7]. On/off timing of activation of each analyzed muscle is calculated from the raw sEMG signal through passband filtering, rectification, envelope computation, and thresholding [8–10], and used to identify the pattern of activation during the gait cycle. Then, the subject-specific pattern can be compared to that of healthy/reference populations for functional assessment [11]. Therefore, the approach used for the identification of such patterns plays a critical role in the resulting functional assessment.

In common practice, a single reference activation pattern is considered, resulting from the analysis of a limited number of strides (i.e., usually few units over a limited number of trials) acquired in laboratory conditions [11]. Recently, the statistical analysis of muscle activity over multiple decades of consecutive strides in ecological conditions highlighted that not only one, but several activation patterns can be identified with different statistical occurrence in the same walking trial, usually one prevalent pattern accompanied by few less frequent, but still statistically significant ones [4,12–16]. Studies performed on healthy adults [4,13,16] found for the rectus femoris a pattern of three activations per gait cycle (i.e., at the beginning of gait cycle, around foot-off, and in the terminal swing) in 53% of total strides, of two activations (i.e., as the previous but lacking activation around foot-off) in 26%, and of two (or three) small activations only around stance-to-swing transition in 17% [13].The identification of this variable behavior is crucial for characterizing motor function and control strategies in healthy subjects, and even more to better understand the modifications occurring in pathological conditions. Therefore, the statistical approach has also been applied in preliminary studies analyzing the muscle activation pattern of pathologic gait; in hemiplegic children, for instance, a reduced frequency in the activation of the tibialis anterior (TA) during terminal swing and a lack of activity of the same muscle at heel-strike compared to healthy controls were identified [17].

Besides the promising preliminary results, and the potential of a better insight in motor control mechanisms provided by the application of a statistical approach for the analysis of sEMG [18,19], some methodological aspects of its implementation still need to be better investigated, especially as referred to the analysis of pathological gait, where changes in the activation pattern can be related to other kinematic and/or kinetic modifications. Gait asymmetry [20,21], for instance, can appear in certain pathological conditions, and can significantly affect the implementation of a statistical approach for the analysis of sEMG, differentiating or not the specific behavior of muscles in the two limbs. Although it is relevant, a specific indication regarding how to take gait symmetry into account is still missing.

Among motor disorders that influence gait performance, Parkinson's Disease (PD) is one of the most investigated in clinical research [22–29]. Usually, authors assess muscle activation during gait of PD patients with amplitude-based parameters [23,24], focusing mainly on distal muscles of the lower limbs, i.e., gastrocnemius (GA) and TA [25–29]. Few recent studies analyzed the on/off timing of these muscles within the gait cycle without identifying a representative activation pattern in PD patients. Cioni et al. [25] observed the absence or extreme reduction in TA activations in early stance or during the early and late swing phases, while Dietz et al. [28] found well preserved timing in the activations of the GA and TA comparable to healthy reference values, and some other authors [27,29] observed the co-activation of the two antagonist muscles (i.e., overlapping periods of on/off timings): Ferrarin et al. [27] reported maintained reciprocity between antagonist muscles, while Dietz et al. [28] observed larger co-activations during the support phase in PD compared to the healthy elderly. These contradictory results can depend on differences in the analyzed PD populations, including different pharmacological treatments and severity of the pathology, but the most important limitation can be identified in the limited number of analyzed strides, preventing the actual identification of significant patterns. Given the intrinsic variability of the muscle activation pattern identified in healthy subjects [10,13], the application of a statistical approach over multiple decades of strides can help to properly take intra-subject variability into account and provide more reliable assessment [30,31] also in pathological gait [17,32].

The aim of the present study was to apply a statistical approach to the analysis of sEMG to identify gastrocnemius medialis (GM) and TA activation patterns in the gait of PD patients, taking potential stride asymmetries into account. More specifically, using data acquired during a 6-min walking test, first, temporal gait symmetry was verified, then occurrence of the different activation patterns, in terms of number and timing within each stride per muscle, and the co-activation of antagonist muscles were quantified.

2. Materials and Methods

2.1. Participants

Twenty PD patients (12 females, 8 males; age 67.2 ± 9.1 years; height 1.65 ± 0.12 m; body mass 67.3 ± 13.1 kg; Hoehn-Yahr stage III, 10 with diagnosis of freezing) participated in the study. All patients were in ON state of Levodopa treatment during the assessment. The study was approved by the local scientific committee and institutional review board (Comitato Etico Interaziendale delle Provincie di Lecco, Como, Sondrio, Italy), and was in accordance with the Code of Ethics of the World Medical Association (Declaration of Helsinki, 1967). A complete explanation of the study protocol was provided to the patients, who signed a written informed consent before their participation in the study. This trial was registered on ClinicalTrials.gov NCT03015714.

2.2. Data Acquisition

Each participant performed a 6-min walking test along a 15 m straight pathway at a self-selected speed wearing own comfortable footwear. Angular velocities of the shanks, used for gait temporal segmentation, were collected using two tri-axial synchronized inertial sensors (OPALs, Apdm, Portland, USA, sampling frequency = 128 Hz) attached to the lateral aspect of each ankle, approximately 0.05 m above the lateral malleolus, using elastic straps.

sEMG signals of GM and TA of both legs were acquired using wireless bipolar surface electromyograph, (Mini Wave, Cometa, Milan, Italy, sampling frequency = 2000 Hz). Sensor placement was performed following SENIAM guidelines [33]: for GM, the electrodes of the sensor were placed on the most prominent bulge of the muscle, aligned longitudinally along the leg [34]; for TA, at 1/3 along the line joining the head of the fibula and the tip of the medial malleolus [34]; bipolar sEMG electrodes are applied around the recommended sensor location with an inter electrode distance of 20 mm.

To synchronize the two systems, the accelerometers embedded in both sEMG sensors (Mini Wave, Cometa, Milan, Italy, sampling frequency = 2000 Hz) and inertial measurement units were used: before acquisition, one of the OPAL sensors was tapped three times on one of the Mini Wave sensor; synchronization was implemented during data processing matching the 3 acceleration peaks acquired by the two systems.

2.3. Data Analysis

Only continuous strides walked along straight paths were considered for the analysis, excluding turns and freezing episodes. Gait events (i.e., foot contact and foot off) were automatically identified from the angular velocity around the medio-lateral axis of the shank [35]. For each subject, stride time was calculated as the difference between two consecutive foot contacts of the same leg.

For the assessment of gait symmetry, the difference between left and right leg of stride time sequences was tested per subject using the Kruskal–Wallis test (statistical significance 5%), since normality of distribution was not verified (Shapiro–Wilk test).

sEMG data were bandpass filtered at 20–450 Hz [36,37], then processed by a double threshold statistical detector to provide the onset and offset time instants of TA and GM activity [38]: per muscle, an amplitude threshold ζ and a numerosity threshold $r0$ were defined; if at least $r0$ out of m consecutive samples, in absolute value, are above ζ, activation is considered on and set to 1; elsewhere activation is considered OFF and set to 0. On-set instants are identified with transitions of the activation from 0 to 1, off-set from 1 to 0.

The behavior of the double-threshold detector is determined by three parameters: the amplitude threshold ζ, the numerosity threshold $r0$, and the length of the observation window m. The values of ζ and $r0$ are statistically selected to minimize the value of false-alarm probability and maximize probability of detection for specific signal-to-noise ratio (SNR) and background noise. To guarantee the performance of the threshold detector, only signals with a minimum SNR value of 10 were considered; the value was chosen according to literature [38]. The values of the background noise level and the SNR were estimated using the statistical approach proposed by Agostini et al. [39]. The length duration of the observation window m was set to 60, i.e., 30 ms, as suitable value for the study of muscle activation in gait analysis [38].

Muscle activation events and intervals (off-set—on-set) per gait cycle were normalized with respect to the corresponding gait cycle duration, then, the number n of times the muscle was activate within a single gait cycle was calculated to define the n-activation pattern per gait cycle and per muscle.

To quantify the frequency of occurrence of each n-activation pattern, muscle activations of each muscle were gathered according to the number of detected intervals within each gait cycle and the occurrence frequency of the single n-activation pattern was calculated per muscle and per subject as:

$$Occurrence\ Frequency\ (n) = \frac{Number\ of\ gait\ cycle\ with\ nactivation\ pattern}{Total\ number\ of\ gait\ cycle}$$

Then, mean and standard dispersion (SD, i.e., standard deviation divided by the square root of the number of strides) values of occurrence frequency of each n-activation patter of each muscle were calculated over subjects.

To characterize the timing of each n-activation pattern, mean and SD of normalized activation events were calculated per n-activation pattern, per muscle, and per patient.

For the co-activation investigation, only patients showing a SNR greater than 10 for both muscles in at least one of the two limbs were considered.

To characterize co-activation [4], per GM n-activation pattern:

- GM activation signal of all gait cycles, normalized with respect to gait cycle duration, were summed up and divided by the number of corresponding gait cycles per subject, then, the resulting subject specific normalized n-activation patterns were summed up and divided by number of subjects.
- The same procedure was replicated per TA n-activation pattern occurring during the same gait cycle of certain GM n-activation pattern.

Coactivation was identified when both concurrent GM and TA normalized n-activation patterns were above 0.1. Matlab R2018a (MathWorks BV, USA) was used for data analysis.

3. Results

Among the 20 subjects, the SNR resulted above the minimum selected value of 10 was in: (i) 5 subjects for both muscles of both legs, (ii) 8 subjects for both muscles of at least for one leg, (iii) 16 subjects for GM and 13 for TA for at least one leg. Therefore, a median of 159 (minimum 48, maximum 208) strides were considered per leg and per subject. For each subject, muscles that showed SNR greater than 10 and considered strides was reported in Table 1.

Table 1. For each subject, the muscles that showed a signal-to-noise ratio (SNR) larger than 10 are indicated with a cross. The number of strides considered per leg are reported in the bottom line.

Subject		**1**	**2**	**3**	**4**	**5**	**6**	**8**	**9**	**10**	**11**	**12**	**13**	**14**	**15**	**16**	**17**	**18**	**19**	**20**
Right	TA			X				X				X			X	X	X			X
	GM	X	X	X	X		X	X						X	X	X		X	X	X
Left	TA	X		X		X	X	X	X		X	X			X	X		X		X
	GM		X	X	X		X	X		X		X	X	X	X	X	X	X	X	X
N. Strides per leg		182	152	179	132	208	64	190	201	194	179	139	48	55	159	203	146	156	170	152

Analyzing gait symmetry: no statistical difference was found between the stride time data series of the left and right leg among subjects (maximum difference for 25th percentile, median and 75th percentile between left and right side for each subject, 0,01 s). Since kinematic symmetry (i.e., stride time) was verified for all subjects, data from left and right leg together were analyzed to calculate n-activation pattern occurrence frequency and timing per subject and per muscle.

For GM, the most frequent (42.9 ± 0.8% occurrence frequency) was the 2-activation pattern: the first activation for all patients at the transition between flat foot contact and push-off; the second activation in 9 out of 16 subjects (1, 3, 6, 8, 12, 13, 14, 16, 19) at the turn of pre- and initial swing, in 2 (17 and 18) during terminal stance, in 1 (4) at initial swing, and in 4 (2, 10, 15, 20) during mid-swing.

The second most frequent (28.0 ± 1.0%) was 1-activation pattern, with a timing similar to the 2-activation pattern for all patients but with no activation during the swing phase.

The third most frequent (21.0 ± 0.9%) was the 3-activation pattern: the first two activations during stance in all subjects but with the exception of 3 (2, 10, 15), around stance-swing transition, in the first half, and at the beginning of swing, respectively; the third during initial and mid-swing for all subjects but 4 (2, 3, 4, 10) during final swing.

The fourth most frequent (7.5 ± 0.6%) was the 4-activation pattern: with an activation scheme similar to the 3-activation pattern during the stance phase and a high variability of the timing of very short activations during the swing phase.

The least frequent (2.5 ± 0.4%) was the 5-activation pattern that, with respect to the other patterns described above, showed several discontinuous activations with high variability during both stance and swing phases, with in some cases (1, 2, 14, and 19) a prolonged activation during mid-stance together with shorter ones.

Occurrence frequency of GM activation pattern is depicted in Figure 1, and timing over gait cycle of the 5 activation patterns for GM in Figure 2.

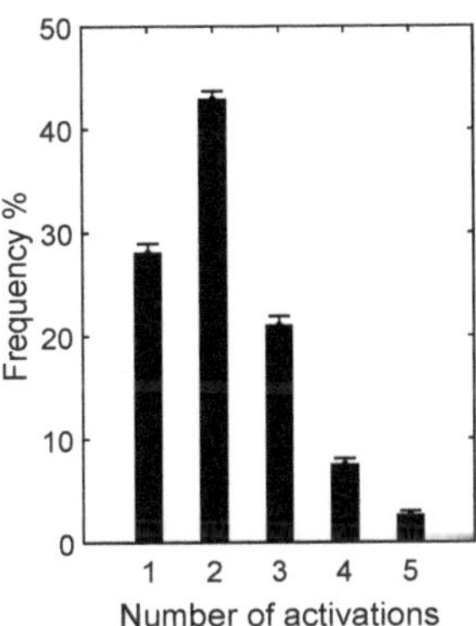

Figure 1. Occurrence frequency (mean ± SD) of the 5 activation patterns of gastrocnemius medialis (GM) over the 16 subjects for whom GM signal-to-noise ratio resulted above the selected threshold of 10.

For TA, the most frequent (39.5 ± 0.7%) was the 3-activation pattern: in 8 subjects out of 13 (1, 3, 6, 11, 15, 16, 17, 18) the first activation at the beginning, and in 5 (5, 8, 9, 12, 20) during mid/final stance; the second in the first half of the swing phase in 6 (3, 5, 6, 8, 15, 18, 20), around the stance to swing-transition in 5 (1, 9, 11, 12, 17), only in 1 subject (16) in the second half of the stance phase; the third in terminal swing for all subjects but 1 (16), anticipating the activation to initial swing phase.

The second most frequent (28.7 ± 0.9%) was the 4-activation pattern: the first at the beginning of stance for all subjects but 1 (20) at mid-stance; the second from mid- to final stance in 9 subjects (1, 3, 6, 9, 11, 12, 16, 17, 18), in 2 (15, 20) around stance to swing transition, and in 2 (5, 8) at the beginning of swing; the third in 5 subjects (5, 6, 15, 18, 20) at mid-swing, in 4 (1, 11, 16, 17) around stance to swing transition, and in 4 (3, 8, 9, 12) in the first half of swing; the fourth at the end of swing in all subjects but 1 (16) at mid-swing.

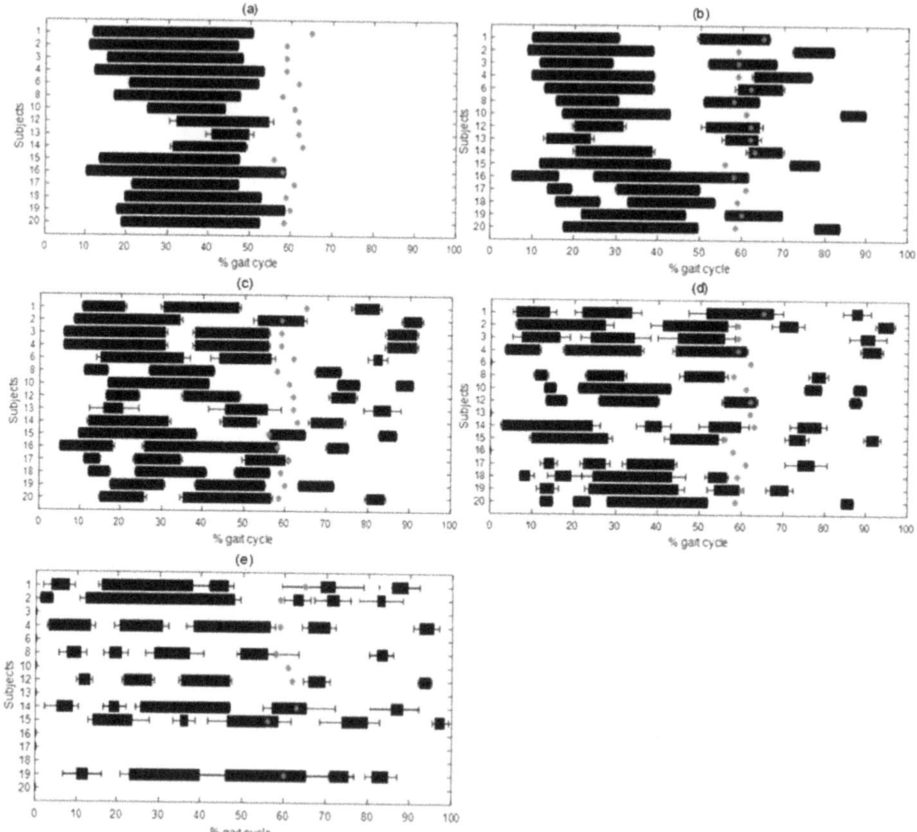

Figure 2. Mean ± SD activation intervals in percentage of gait cycle for 1- (**a**), 2- (**b**), 3- (**c**), 4- (**d**) and 5- (**e**) activation pattern in GM. Mean value of foot-off in percentage of gait cycle is represented with a red star, representing the beginning of swing phase.

The third most frequent (17.9 ± 1.0%) was the 2-activation pattern: the first in 5 subjects (1, 9, 11, 12, 18) during the second half of stance, in 4 (3, 5, 8, 17) around stand to swing transition, in 2 (6, 16) during the first half pf stance, in 1 (15) at midstance, and in 1 (20) during the first half of the swing phase; the between mid- or final swing for all subjects but 2 (6, 16) during initial swing.

The fourth (9.8 ± 0.6%) was the 5-activation pattern: first activation at the beginning of stance for all subjects; second variably distributed from initial to late stance in all subjects; third in 6 subjects (1, 9, 11, 12, 16, 17) in the second half of stance, in 2 (3, 8) at the stance to swing transition, in 5 (5, 6, 15, 18, 20) at the beginning of swing; the fourth between initial and mid-swing for all subjects but 2 around stance to swing transition (1) and in late swing (6); the fifth at the end of swing in 10 subjects (1, 3, 6, 8, 9, 11, 15, 17, 18, 20), at mid-swing in 2 (5, 16), and at early swing in 1 (12).

The least frequent (3.0 ± 0.4%) was the 1-activation pattern, exhibited only by 6 out of 13 subjects (1, 3, 5, 9, 12, 18), varying largely between the end of stance and the end of swing.

Occurrence frequency of TA activation pattern is depicted in Figure 3, and timing over gait cycle of the five activation patterns for TA in Figure 4.

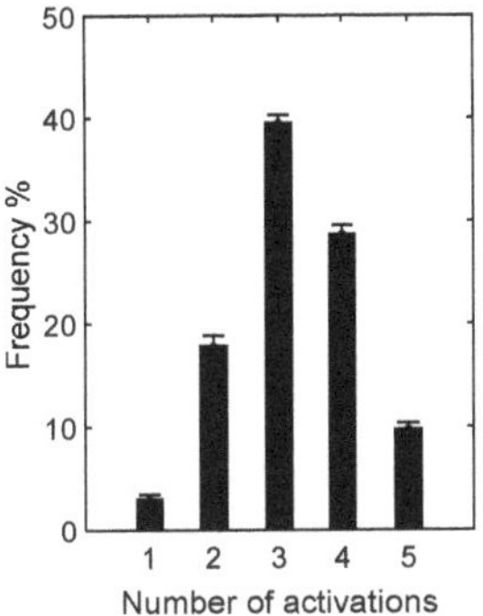

Figure 3. Occurrence frequency (mean ± SD) of the 5 activation patterns of tibialis anterior (TA) over the 13 subjects for whom TA signal-to-noise ratio resulted above the selected threshold of 10.

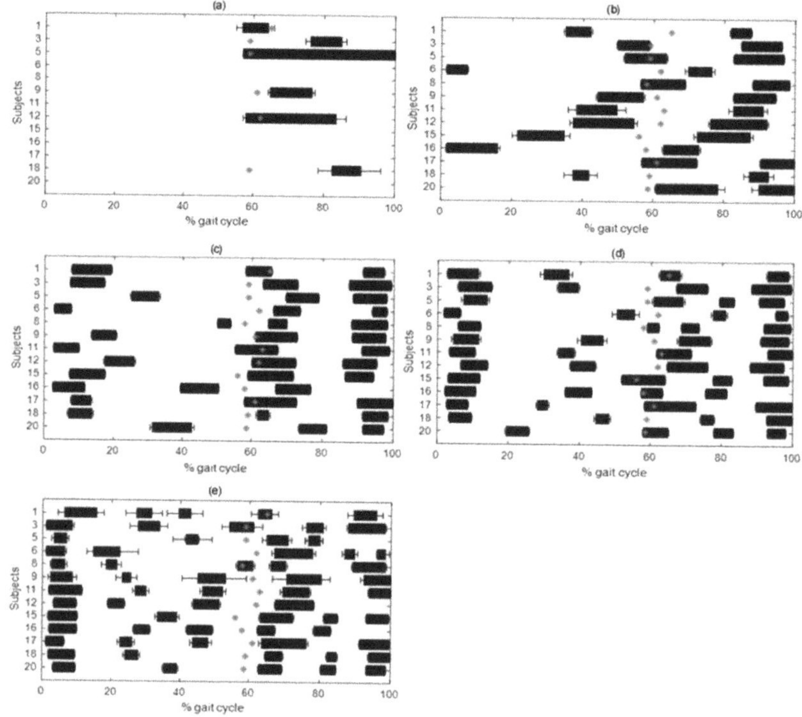

Figure 4. Mean ± SD activation intervals in percentage of gait cycle for 1- (**a**), 2- (**b**), 3- (**c**), 4- (**d**), and 5- (**e**) activation pattern in TA. Mean value of foot-off in percentage of gait cycle is represented with a red star, representing the beginning of swing phase.

All GM n-activation patterns showed TA coactivation, for TA 2-, 3-, 4-activation patterns, at the beginning of stance from foot contact, approximately from 0% to 20% of the gait cycle, and in preparation of foot off, approximately from 55% to 65% of the gait cycle. With more than 1 GM activation (i.e., 2, 3, 4-activation patterns) a third TA coactivation phase was identified from mid-swing to foot contact again, approximately from 75% to 100% of the gait cycle.

Coactivation for TA 1-activation and 5-activation patterns was not reported due to rare occurrence of the former and large variability of intermediate activations in the latter.

Normalized n-activation patterns of GM and TA and corresponding coactivation intervals are depicted in Figure 5.

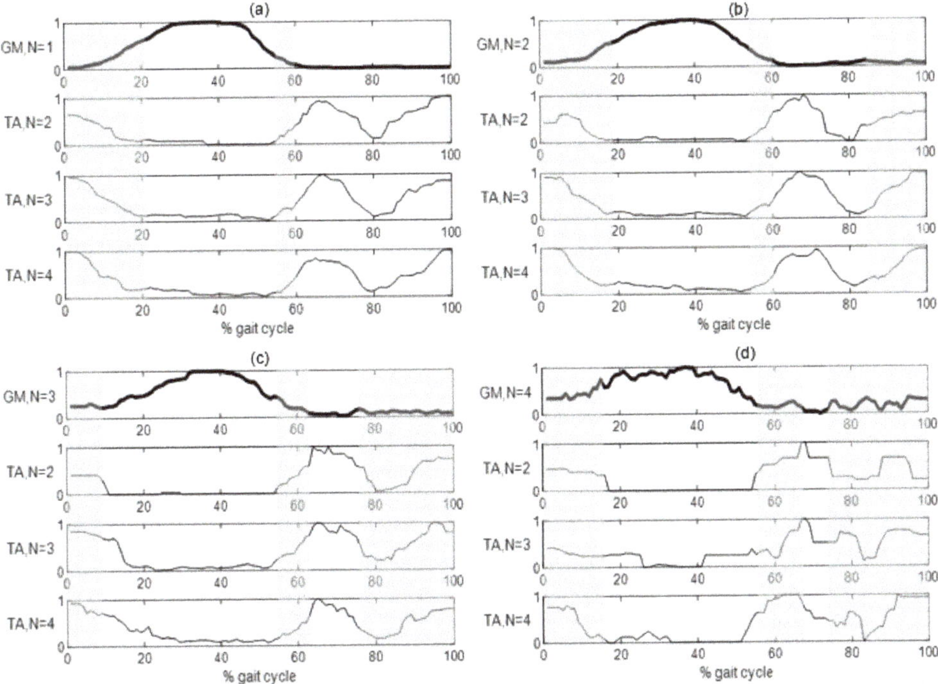

Figure 5. Normalized activation intervals of TA detected strides where GM (thicker lines) showed 1- (**a**), 2- (**b**), 3- (**c**) and 4- (**d**) activation pattern. Co-contractions intervals, where both concurrent patterns are above 0.1, are highlighted by vertical grey bands.

4. Discussion

The present work analyzed sEMG in a population of 20 PD patients for the assessment of GM an TA activation patterns during gait using a statistical approach for the processing of a large number of strides.

Previous studies [25–29,40] evaluating muscle activation in PD patients reported non-concordant results probably due to the limited (from few units to a couple of tens) number of gait cycles per subject, indicating a single reference activation pattern for each muscle.

The relatively recent application of a statistical approach for the analysis of sEMG from long gait sequences (i.e., multiple decades of gait cycles) highlighted that different activation patterns can occur even in healthy gait [4,12–16] and in hemiplegic children [32], as preliminary application to pathological conditions. In the present study, the application of a statistical approach was applied in PD patients to long gait sequences from a 6-min walking test (i.e., in the order of multiple decades of gait cycles) to identify the different possible gait patterns.

Approaching the analysis of gait in pathology, specific conditions have to be addressed. Previous studies that applied statistical EMG analysis to the gait of healthy subjects did not tackle the issue of gait symmetry, as in healthy subjects no difference between left and right sides is reasonably considered [15,16]. Considering the pathological population under investigation in the present study, this hypothesis must be verified exploiting an approach coherent with the further statistical analysis on the on/off timing activation pattern. In the present study, overall symmetry was investigated using stride time as a discrete temporal indicator in accordance with previous studies [20].

The application of the statistical approach [38] to sEMG in the analyzed PD population after gait symmetry assessment allowed us to identify different activation patterns for both GM and TA, supporting the existence of an intrinsic variability in EMG activation schemes even in a paradigmatic motor task as gait, as already observed in healthy subjects [10,12,17] and hemiplegic children [32]. In more detail, 5 activation patterns were identified by the computational procedure for both muscles, although, based on the quantification of occurrence frequency, only four can be considered significant for GM (i.e., the 5-activation pattern excluded having an occurrence frequency below 3%), and four for TA (i.e., the 1-activation pattern excluded due to occurrence frequency below 4% and in a minority of subjects), although the TA 5-activation pattern presented a low but significant occurrence frequency (in the order of 10%), resulted too variable in timing to allow a systematic characterization.

As already observed in healthy subjects, GM activity was centered between flat foot contact and push off, and in the final swing: even if shorter and delayed, the first activation is also found in PD, highlighting the active participation in stabilizing ankle dorsiflexion during the forward progression [3,11]; the second region of activity mainly occurred during the early- and mid-swing, anticipating the activation timing observed in healthy adults [12]. When no statistical approach was applied and only the triceps surae mean single timing pattern was estimated, this second region of activity was not observed [31]. For TA, the muscular activity of healthy subjects resulted to be centered from the pre-swing to the following loading response and during the mid-stance [12]. In the analyzed PD subjects, the activity at the beginning of the gait cycle was absent or delayed, showing alterations that normally contribute to foot positioning at the touch down [27]. Moreover, the activations during the swing phase were generally shorter and anticipated, showing a compensation of the anticipation in the activation of GM during the swing phase [27]. These differences in the TA activation pattern were not observed previously with the mean single pattern approach [31].

The coactivation analysis of the two antagonist muscles, GM vs TA, was applied to understand the mechanisms of regulation of joint stiffness and stability [41,42]. To the authors' knowledge, only one study assessed the coactivation of antagonist muscles using statistical approaches, investigating GA and quadriceps femoris during gait of healthy adults [15]. Following the same methodological approach [15], the co-contraction between GM and TA was quantified as the overlapping of activation intervals. In the analyzed PD subjects, results showed two or three GM-TA co-activation intervals within the single gait cycle (Figure 5), confirming the maintained reciprocity of antagonist nature of considered muscles [27]: the first coactivation lasted from heel strike to mid-stance and can be associated to ankle stabilization during the early stance [43]; the second activation occurred from the pre—to initial swing, when TA contributes to rapid dorsi-flexion of the foot during swing phase [44], while GM activity is commonly related to the plantar-flexion needed for the heel raising [45]; the third and last coactivation occurred during late swing, when TA acts as ankle dorsi-flexor and GM as foot-invertor [12], to properly positioning and stabilizing the foot for the following foot contact [11,46]. First and second coactivations occurred for every GM and TA n-activation pattern, while the third coactivation did not occur for the GM 1-activation pattern, independently from the associated TA n-activation pattern.

Due to the number of analyzed PD subjects and their specific characteristics (e.g., severity and ON state of Levodopa therapy), the results of the present study cannot be generalized to PD population; nevertheless, the exploited statistical approach, applied to a large number of strides acquired in ecological conditions, demonstrated the capability to also identify different activation patterns in the target pathologic population, allowing to quantify the occurrence frequency of each, analyze activation timing distribution per activation pattern, and coactivation as related to specific activation patterns, supporting the methodological aim of the present study.

The use of a statistical approach to identify activation patterns from sEMG in the selected pathologic population supported a better insight in the analysis of muscle activation and control, and can be used for monitoring and understanding how muscle activity changes with the development of the pathology. Moreover, it provided quantitative parameters to support further analysis (e.g., occurrence frequency of each n-activation pattern, presence/absence of a certain muscle activity in a specific gait phase) that can also be exploited to size sample and gait duration or can be included, together with spatio-temporal parameters, in the statistical methods/models for the general analysis of gait in PD patients [47–49].

In addition, the approach allows for the estimate of SNR on the signal, and the constraint on SNR allows us to guarantee the quality of the analyzed data and of the resulting outcome, although the constraint can lead to the discarding of part of the acquired data, as not all gait cycles can be included in the analysis (Table 1). Therefore, for the application of statistical approach, not only do long sequences of gait have to be acquired, but also great care must be paid during the experimental procedure (e.g., skin preparation, electrode placement etc.) to minimize the loss of acquired data. The verification of symmetry, as in the present study, allows us to increment the statistical power of the data-set, computing data from the right and left side together, nevertheless, this verification is not guaranteed in pathologic subjects and specific inspection must be previously performed.

5. Conclusions

The present study evaluated GM and TA activation in PD patients, proposing a comprehensive step-to-step analysis procedure based on a statistical approach from the detection of activation patterns from long gait sequences acquired in ecological conditions, integrated with verification of gait symmetry and coactivation analysis. The analysis allowed to identify different muscle activation patterns and their relative occurrence frequency, as well as coactivation schemes, in PD subjects, providing novel insight with respect to literature, where unique average patterns were identified, providing at times conflicting results. The improved methodology, here applied as an example to GM and TA in 20 PD subjects, is meant to be considered as a general procedure to be applied for the analysis of muscle activation patterns in pathologic subjects.

Author Contributions: Data curation, G.P.P., M.F.; Formal analysis, S.F. and R.S.; Methodology, G.P.P., S.F., R.S.; Project administration, D.F., G.F., S.F., R.S.; Software, G.P.P.; Supervision, S.F., R.S.; Writing—original draft, G.P.P.; Writing—review and editing, G.P.P., M.C.B., S.F., R.S. All authors have read and agreed to the published version of the manuscript.

Funding: This research received no external funding.

Conflicts of Interest: The authors declare no conflict of interest.

References

1. Mortka, K.; Wiertel-Krawczuk, A.; Lisiński, P. Muscle Activity Detectors—Surface Electromyography in the Evaluation of Abductor Hallucis Muscle. *Sensors* **2020**, *20*, 2162. [CrossRef] [PubMed]
2. Türker, H.; Sze, H.; Sözen, H.T.A.H. Surface Electromyography in Sports and Exercise. *Electrodiagnosis New Front. Clin. Res.* **2013**. [CrossRef]
3. Sutherland, D. The evolution of clinical gait analysis part l: Kinesiological EMG. *Gait Posture* **2001**, *14*, 61–70. [CrossRef]
4. Di Nardo, F.; Agostini, V.; Knaflitz, M.; Mengarelli, A.; Maranesi, E.; Burattini, L.; Fioretti, S. The occurrence frequency: A suitable parameter for the evaluation of the myoelectric activity during walking. In Proceedings of the 2015 37th Annual International Conference of the IEEE Engineering in Medicine and Biology Society (EMBC), Milan, Italy, 25–29 August 2015; pp. 6070–6073. [CrossRef]
5. Zhang, X.; Zhou, P. Sample entropy analysis of surface EMG for improved muscle activity onset detection against spurious background spikes. *J. Electromyogr. Kinesiol. Off. J. Int. Soc. Electrophysiol. Kinesiol.* **2012**, *22*, 901–907. [CrossRef]
6. Liu, J.; Ying, D.; Rymer, W.Z.; Zhou, P. Robust Muscle Activity Onset Detection Using an Unsupervised Electromyogram Learning Framework. *PLoS ONE* **2015**, *10*, e127990. [CrossRef]
7. De Luca, C.J. The Use of Surface Electromyography in Biomechanics. *J. Appl. Biomech.* **1997**, *13*, 135–163. [CrossRef]
8. Tirosh, O.; Sangeux, M.; Wong, M.; Thomason, P.; Graham, K. Walking speed effects on the lower limb electromyographic variability of healthy children aged 7–16 years. *J. Electromyogr. Kinesiol.* **2013**, *23*, 1451–1459. [CrossRef]

9. Chang, W.N.; Lipton, J.S.; Tsirikos, A.I.; Miller, F. Kinesiological surface electromyography in normal children: Range of normal activity and pattern analysis. *J. Electromyogr. Kinesiol.* **2007**, *17*, 437–445. [CrossRef]
10. Agostini, V.; Nascimbeni, A.; Gaffuri, A.; Imazio, P.; Benedetti, M.G.; Knaflitz, M. Normative EMG activation patterns of school-age children during gait. *Gait Posture* **2010**, *32*, 285–289. [CrossRef]
11. Perry, J.; Burnfield, J. *Gait Analysis: Normal and Pathological Function*, 2nd ed.; Slack Inc: Thorofare, NJ, USA, 2010.
12. Di Nardo, F.; Ghetti, G.; Fioretti, S. Assessment of the activation modalities of gastrocnemius lateralis and tibialis anterior during gait: A statistical analysis. *J. Electromyogr. Kinesiol. Off. J. Int. Soc. Electrophysiol. Kinesiol.* **2013**, *23*, 1428–1433. [CrossRef]
13. Di Nardo, F.; Fioretti, S. Statistical analysis of surface electromyographic signal for the assessment of rectus femoris modalities of activation during gait. *J. Electromyogr. Kinesiol.* **2013**, *23*, 56–61. [CrossRef] [PubMed]
14. Di Nardo, F.; Maranesi, E.; Mengarelli, A.; Ghetti, G.; Burattini, L.; Fioretti, S. Assessment of the variability of vastii myoelectric activity in young healthy females during walking: A statistical gait analysis. *J. Electromyogr. Kinesiol.* **2015**, *25*, 800–807. [CrossRef] [PubMed]
15. Mengarelli, A.; Gentili, A.; Strazza, A.; Burattini, L.; Fioretti, S.; Di Nardo, F. Co-activation patterns of gastrocnemius and quadriceps femoris in controlling the knee joint during walking. *J. Electromyogr. Kinesiol.* **2018**, *42*, 117–122. [CrossRef] [PubMed]
16. Di Nardo, F.; Mengarelli, A.; Maranesi, E.; Burattini, L.; Fioretti, S. Gender differences in the myoelectric activity of lower limb muscles in young healthy subjects during walking. *Biomed. Signal Process. Control* **2015**, *19*, 14–22. [CrossRef]
17. Di Nardo, F.; Strazza, A.; Mengarelli, A.; Cardarelli, S.; Tigrini, A.; Verdini, F.; Nascimbeni, A.; Agostini, V.; Knaflitz, M.; Fioretti, S. EMG-Based Characterization of Walking Asymmetry in Children with Mild Hemiplegic Cerebral Palsy. *Biosensors* **2019**, *9*, 82. [CrossRef]
18. Grazia, M.; Knaflitz, M.; Agostini, V.; Bonato, P. Muscle Activation Patterns During Level Walking and Stair Ambulation. *Appl. EMG Clin. Sports Med.* **2012**. [CrossRef]
19. Conti, M.; Martínez Madrid, N.; Seepold, R.; Orcioni, S. *Mobile Networks for Biometric Data Analysis*; Springer International Publishing: Cham, Switzerland, 2016. [CrossRef]
20. Vitečková, S.; Kutilek, P.; Svoboda, Z.; Krupička, R.; Kauler, J.; Szabó, Z. Gait symmetry measures: A review of current and prospective methods. *Biomed. Signal Process. Control* **2018**, *42*, 89–100. [CrossRef]
21. Moevus, A.; Mignotte, M.; De Guise, J.A.; Meunier, J. Evaluating perceptual maps of asymmetries for gait symmetry quantification and pathology detection. In Proceedings of the 2014 36th Annual International Conference of the IEEE Engineering in Medicine and Biology Society, Chicago, IL, USA, 26–30 August 2014; pp. 3317–3320. [CrossRef]
22. Patikas, D.A. EMG Activity in Gait: The Influence of Motor Disorders. In *Handbook of Human Motion*; Müller, B., Wolf, S.I., Brueggermann, G.P., Deng, Z., McIntosh, A., Miller, F., Selbie, W.S., Eds.; Springer International Publishing: Cham, Switzerland, 2016; pp. 1–26.
23. Bailey, C.A.; Corona, F.; Murgia, M.; Pili, R.; Pau, M.; Côté, J.N. Electromyographical Gait Characteristics in Parkinson's Disease: Effects of Combined Physical Therapy and Rhythmic Auditory Stimulation. *Front. Neurol.* **2018**, *9*. [CrossRef]
24. Nieuwboer, P.A.; Dom, R.; De Weerdt, W.; Desloovere, K.; Janssens, L.; Stijn, V. Electromyographic profiles of gait prior to onset of freezing episodes in patients with Parkinson's disease. *Brain* **2004**, *127*, 1650–1660. [CrossRef]
25. Cioni, M.; Richards, C.L.; Malouin, F.; Bedard, P.J.; Lemieux, R. Characteristics of the electromyographic patterns of lower limb muscles during gait in patients with Parkinson's disease when OFF and ON L-Dopa treatment. *Ital. J. Neurol. Sci.* **1997**, *18*, 195–208. [CrossRef]
26. Mitoma, H.; Hayashi, R.; Yanagisawa, N.; Tsukagoshi, H. Characteristics of parkinsonian and ataxic gaits: A study using surface electromyograms, angular displacements and floor reaction forces. *J. Neurol. Sci.* **2000**, *174*, 22–39. [CrossRef]
27. Ferrarin, M.; Carpinella, I.; Rabuffetti, M.; Rizzone, M.; Lopiano, L.; Crenna, P. Unilateral and Bilateral Subthalamic Nucleus Stimulation in Parkinson's Disease: Effects on EMG Signals of Lower Limb Muscles During Walking. *IEEE Trans. Neural Syst. Rehabil. Eng.* **2007**, *15*, 182–189. [CrossRef] [PubMed]

28. Dietz, V.; Zijlstra, W.; Prokop, T.; Berger, W. Leg muscle activation during gait in Parkinson's disease: Adaptation and interlimb coordination. *Electroencephalogr. Clin. Neurophysiol. Mot. Control.* **1995**, *97*, 408–415. [CrossRef]
29. Dietz, V.; Quintern, J.; Berger, W. Electrophysiological Studies of Gait in Spasticity and Rigidity Evidence that Altered Mechanical Properties of Muscle Contribute to Hyperonia. *Brain* **1981**, *104*, 431–449. [CrossRef] [PubMed]
30. Agostini, V.; Balestra, G.; Knaflitz, M. Segmentation and Classification of Gait Cycles. *IEEE Trans. Neural Syst. Rehabil. Eng.* **2014**, *22*, 946–952. [CrossRef]
31. Rosati, S.; Agostini, V.; Knaflitz, M.; Balestra, G. Muscle activation patterns during gait: A hierarchical clustering analysis. *Biomed. Signal Process. Control* **2017**, *31*, 463–469. [CrossRef]
32. Agostini, V.; Nascimbeni, A.; Gaffuri, A.; Knaflitz, M. Multiple gait patterns within the same Winters class in children with hemiplegic cerebral palsy. *Clin. Biomech.* **2015**, *30*, 908–914. [CrossRef]
33. Hermens, H.; Freriks, B. *The State of the Art on Sensors and Sensor Placement Procedures for Surface Electromyography: A Proposal for Sensor Placement Procedures*; Roessingh Research and Development: Enschede, The Netherlands, 1997.
34. Welcome to SENIAM. Available online: http://www.seniam.org/ (accessed on 16 September 2020).
35. Salarian, A.; Russmann, H.; Vingerhoets, F.; Dehollain, C.; Blanc, Y.; Burkhard, P.; Aminian, K. Gait Assessment in Parkinson's Disease: Toward an Ambulatory System for Long-Term Monitoring. *IEEE Trans. Biomed. Eng.* **2004**, *51*, 1434–1443. [CrossRef]
36. Liu, J.; Ying, D.; Rymer, W.Z. EMG burst presence probability: A joint time-frequency representation of muscle activity and its application to onset detection. *J. Biomech.* **2015**, *48*, 1193–1197. [CrossRef]
37. Rashid, U.; Niazi, I.K.; Signal, N.E.J.; Farina, D.; Taylor, D. Optimal automatic detection of muscle activation intervals. *J. Electromyogr. Kinesiol.* **2019**, *48*, 103–111. [CrossRef]
38. Bonato, P.; D'Alessio, T.; Knaflitz, M. A statistical method for the measurement of muscle activation intervals from surface myoelectric signal during gait. *IEEE Trans. Biomed. Eng.* **1998**, *45*, 287–299. [CrossRef] [PubMed]
39. Agostini, V.; Knaflitz, M. An Algorithm for the Estimation of the Signal-To-Noise Ratio in Surface Myoelectric Signals Generated During Cyclic Movements. *IEEE Trans. Biomed. Eng.* **2012**, *59*, 219–225. [CrossRef] [PubMed]
40. Albani, G.; Sandrini, G.; Künig, G.; Martin-Soelch, C.; Mauro, A.; Pignatti, R.; Pacchetti, C.; Dietz, V.; Leenders, K.L. Differences in the EMG pattern of leg muscle activation during locomotion in Parkinson's disease. *Funct. Neurol.* **2003**, *18*, 165–170. [PubMed]
41. Schmitz, A.; Silder, A.; Heiderscheit, B.; Mahoney, J.; Thelen, D.G. Differences in lower-extremity muscular activation during walking between healthy older and young adults. *J. Electromyogr. Kinesiol. Off. J. Int. Soc. Electrophysiol. Kinesiol.* **2009**, *19*, 1085–1091. [CrossRef]
42. Hortobágyi, T.; DeVita, P. Muscle pre- and coactivity during downward stepping are associated with leg stiffness in aging. *J. Electromyogr. Kinesiol.* **2000**, *10*, 117–126. [CrossRef]
43. Cardarelli, S.; Gentili, A.; Mengarelli, A.; Verdini, F.; Fioretti, S.; Burattini, L.; Di Nardo, F. Ankle Muscles Co-Activation Patterns During Normal Gait: An Amplitude Evaluation. In *EMBEC & NBC 2017. EMBEC 2017, NBC 2017. IFMBE Proceedings*; Eskola, H., Väisänen, O., Viik, J., Hyttinen, J., Eds.; Springer: Singapore, 2017; Volume 65. [CrossRef]
44. Lichtwark, G. The role of the tibialis anterior muscle and tendon in absorbing energy during walking. *J. Sci. Med. Sport* **2014**, *18*, e129. [CrossRef]
45. Winter, D.A. *The Biomechanics and Motor Control of Human Gait*; University of Waterloo Press: Waterloo, ON, Canada, 1987.
46. Mengarelli, A.; Maranesi, E.; Burattini, L.; Fioretti, S.; Di Nardo, F. Co-contraction activity of ankle muscles during walking: A gender comparison. *Biomed. Signal Process. Control* **2017**, *33*, 1–9. [CrossRef]
47. Wu, Y.; Krishnan, S. Statistical Analysis of Gait Rhythm in Patients with Parkinson's Disease. *IEEE Trans. Neural Syst. Rehabil. Eng. Publ. IEEE Eng. Med. Biol. Soc.* **2010**, *18*, 150–158. [CrossRef]

48. Khajuria, A.; Joshi, P.; Joshi, D. Comprehensive Statistical Analysis of the Gait Parameters in Neurodegenerative Diseases. *Neurophysiology* **2018**, *50*, 38–51. [CrossRef]
49. Arcolin, I.; Corna, S.; Giardini, M.; Giordano, A.; Nardone, A.; Godi, M. Proposal of a new conceptual gait model for patients with Parkinson's disease based on factor analysis. *Biomed. Eng. Online* **2019**, *18*, 70. [CrossRef]

Article

Variability of Muscular Recruitment in Hemiplegic Walking Assessed by EMG Analysis

Francesco Di Nardo [1,*], Susanna Spinsante [1], Chiara Pagliuca [1], Angelica Poli [1], Annachiara Strazza [1], Valentina Agostini [2], Marco Knaflitz [2] and Sandro Fioretti [1,*]

1 Department of Information Engineering, Università Politecnica delle Marche, 60100 Ancona, Italy; s.spinsante@staff.univpm.it (S.S.); chiara_pagliuca@yahoo.it (C.P.); angelicapoli93@gmail.com (A.P.); a.strazza@pm.univpm.it (A.S.)
2 Department of Electronics and Telecommunications, Politecnico di Torino, 10129 Torino, Italy; valentina.agostini@polito.it (V.A.); marco.knaflitz@polito.it (M.K.)
* Correspondence: f.dinardo@staff.univpm.it (F.D.N.); s.fioretti@staff.univpm.it (S.F.); Tel.: +39-071-220-4838 (F.D.N.)

Received: 26 August 2020; Accepted: 23 September 2020; Published: 25 September 2020

Abstract: Adaptive variability during walking is typical of child motor development. It has been reported that neurological disorders could affect this physiological phenomenon. The present work is designed to assess the adaptive variability of muscular recruitment during hemiplegic walking and to detect possible changes compared to control populations. In the attempt of limiting the complexity of computational procedure, the easy-to-measure coefficient of variation (CV) index is adopted to assess surface electromyography (sEMG) variability. The target population includes 34 Winters' type I and II hemiplegic children (H-group). Two further healthy populations, 34 age-matched children (C-group) and 34 young adults (A-group), are involved as controls. Results show a significant decrease ($p < 0.05$) of mean CV for gastrocnemius lateralis (GL) in H-group compared to both C-group (15% reduction) and A-group (35% reduction). Reductions of mean CV are detected also for tibialis anterior (TA) in H-group compared to C-group (7% reduction, $p > 0.05$) and A-group (15% reduction, $p < 0.05$). Lower CVs indicate a decreased intra-subject variability of ankle-muscle activity compared to controls. Novel contribution of the study is twofold: (1) To propose a CV-based approach for an easy-to-compute assessment of sEMG variability in hemiplegic children, useful in different experimental environments and different clinical purposes; (2) to provide a quantitative assessment of the reduction of intra-subject variability of ankle-muscle activity in mild-hemiplegic children compared to controls (children and adults), suggesting that hemiplegic children present a limited capability of adapting their muscle recruitment to the different stimuli met during walking task. This finding could be very useful in deepening the knowledge of this neurological disorder.

Keywords: surface electromyography; cerebral palsy; hemiplegia; motor disorders; gait variability; coefficient of variation

1. Introduction

Hemiplegia, often observed in children affected by cerebral palsy, is a neurological disease characterized by the fact that only half of the body is affected by the disorder. Modified selective motor control, weakness and spasticity are associated with hemiplegia, conditioning everyday activities including walking [1]. In the late 1980s, Winters et al. introduced a suitable classification of gait in hemiplegia. Based on a kinematics analysis, the authors identified four different gait patterns in the sagittal plane where four categories were discriminated, based on a progressive distal-proximal involvement of the hemiplegic leg [2]. Winters' type I patients present a hypo-activation of dorsi-flexor muscles of the hemiplegic-leg ankle, causing drop foot during swing. Winters' type II subjects are typified by the persistence of equinism throughout the gait cycle, often related to a hyperextension of the knee during stance phase [2,3]. Winters' type I and II are the forms of hemiplegia most frequently detected in cerebral palsy; thus, many studies have focused on them [3–6].

Surface electromyography (sEMG) is an acknowledged diagnostic technique, typically used to characterize muscular activity by means of a non-invasive approach. Non-invasiveness and easiness of use, associated with the increasing availability of solutions based on sEMG, make this technique particularly valuable for the analysis of those pathologies in which walking is directly affected, as in cerebral palsy. Many studies, including our own, used an sEMG-based approach to identify the gait patterns adopted by hemiplegic children and to compare them with control children [4,6–11]. In particular, two recent studies, performed on numerous strides (hundreds) per patient, reported clear alterations of muscular-recruitment patterns in hemiplegic side: Reduced and less frequent activity during swing and a dearth of activity at loading response of tibialis anterior in type I and II; and a hyper-activation of gastrocnemius around initial contact was identified in type II only [5,6]. Overall, both studies reported significant variability in activation modality of muscles of both hemiplegic leg [5] and contralateral (non-hemiplegic) one [6].

Adaptive variability is typical of human motor development. According to some researchers, the variability in early infant movements is a key aspect of motor development [12]. Moreover, other sEMG-based studies suggest that an initial attempt of adaptation in postural behavior during sitting could be identified in four-month-old infants [13]. Then, all basic motor functions will achieve the first stages of the so-called secondary variability around the age of 18 months. Active trial-and-error experiences, specific to each subject, are typical of this stage. The basic, variable motor skill reached during the phase of primary variability keeps on developing and modifying all through the subject's life, allowing increasingly accurate and organized movements. Consequently, adult subjects master a wide movement repertoire, enabling an efficient motor solution for each specific circumstance [13]. Overall variability of human motion is associated with variability of muscle activity, quantified by EMG signals. In a preliminary study of the present group of researchers [14], sEMG-signal variability was quantified in relation to motor development, comparing adult and children populations by means of a quantitative index, the coefficient of variation (CV), previously tested on different EMG signals [14–16]. That study suggested that CV is an easy-to-measure index able to quantify sEMG variability in different experimental conditions and with different clinical purposes: In adult and pediatric populations and for both intra- and inter-subject studies. sEMG variability has been infrequently assessed in hemiplegic children and only by means of computationally expensive techniques, such as statistical gait analysis [4,6]. To the authors' best knowledge, the CV index has never been applied to quantify the variability of muscular recruitment during hemiplegic walking. Moreover, no attempts were reported in literature to provide a direct and quantitative comparison of sEMG-variability values between hemiplegic children and controls.

Thus, the present work is designed to assess the adaptive variability of muscular recruitment during hemiplegic walking and to detect possible changes in sEMG variability of hemiplegic walking compared to controls. The easy-to-measure CV index is chosen to achieve this goal, in order to propose a novel approach able to limit the complexity of computational procedures. The CV value is computed in 34 school-age hemiplegic children identified as type I and II by Winters' classification and in a large number of cycles per subject (hundreds), resulting in around 30,000 strides in total, to guarantee an adequate number of samples for variability characterization. The same index is used to describe sEMG behavior in two further populations, school-age children (34 subjects) and young adults (34 subjects), to compare and interpret results achieved in the hemiplegic population. The manuscript is organized as follows: Section 2 provides a short summary of the main indices available in the literature and used to quantify and analyze the sEMG signal variability in different scenarios, among which the CV is applied in the present study. Section 3 presents material and methods based on which the research was developed, providing details about sEMG processing, test populations and parameters computation. Section 4 presents the experimental results that are discussed in Section 5, along with retrospection on the related state of the art. Finally, Section 6 concludes the manuscript and provides insights for future research developments.

2. Indices for sEMG Variability Analysis

The non-invasive recording of muscle electrical activity during dynamic tasks is greatly supported by sEMG, thanks to a huge collection of algorithms and techniques specifically designed to obtain and interpret the muscle activation patterns. The last ones may appear in patients with altered locomotion, and the use of sEMG in clinical gait analysis helps identifying such a condition. Despite the aforementioned advantages, and the market availability of wireless, lightweight and minimally invasive sEMG measurement equipment, such as the Myon [17] or the Freeemg [18] devices, sEMG has not witnessed a pervasive and widespread adoption in clinical assessment or rehabilitation yet. This is motivated by education barrier, i.e., understanding the features and information associated with electrical signals measured on the body may be not easy or straightforward by clinical operators [19,20]. Additional complexity is determined by the possibility to apply a huge variety of parameters, indices and figures, differently defined and computed from the measured sEMG signal samples, according to the specific muscle feature or activation pattern one is interested to observe [21]. For example, root mean square (RMS), median frequency (MF) and mean power frequencies (MPF) based on Fourier Transform [22] have been effectively used in applications dealing with the evaluation of muscular fatigue.

It is well recognized that the human motor system exhibits redundancy, so a single motor task may be performed in several different ways, leading to a similar final result [23]. Redundancy of motor repertoire in human subjects reflects the capability of the nervous system to generate different patterns of muscle activation, for the same given movement. Such a capability motivates either intra- and inter-subject variability of muscle activation, which can be captured by suitably designed indices computed on the measured sEMG signal samples. For example, indices proposed for sEMG analysis focused on aspects pertaining to running are mean, standard deviation (SD) and mean CV, as well as CV calculated over the running cycle [24]. The mean sEMG value at the denominator of the CV definition influences the value of such an index: For sensors located in those body areas where muscle activity is very weak or not present at all, the variability may be overestimated [25]. In order to overcome this limit, other metrics have been introduced, such as the variance ratio (VR) applied in gait analysis [26]. In studying intra-individual variability of sEMG in front crawl swimming, Martens et al. [27] introduced several one- and two-dimension metrics: They included both one- and two-dimension CV, VR and the coefficient of quartile variation (CQV). Corresponding general definitions are reported in Table 1. In particular, the CV of a quantity is defined as the ratio of its standard deviation to its mean, as given in Table 1. Such an index is largely used in many clinical fields, but it is not commonly applied to sEMG signals. In the present work, CV is adopted to quantify the variability of muscle rhythmic activation during walking in three different populations, namely hemiplegic children, healthy school children and young adult. Motivation for choosing the CV is threefold: (i) We aim for applying and testing this index in the evaluation of sEMG variability during walking in hemiplegic children for the first time at our best knowledge; (ii) we aim for checking the suitability of such an easy-to-compute index in reflecting different characteristics between pathological and control children and then between children and young adults, in order to promote the adoption of sEMG in clinical practice: Despite its simplicity, the index is able to satisfactorily discriminate the muscular recruitment during walking exhibited by different populations [14,28]; (iii) CV is a unit-free measure, suitable to compare normally distributed data by directly quantifying the degree of variability relative to the mean of the distributions [28]. The CV index, indeed, is not directly computed on sEMG samples, but it is derived from the standard deviation of the signal, which is by definition a direct measurement of the signal variability. These characteristics seem to make this index more suitable to the aim of the present study, respect to CQV and VR indices. CQV index, indeed, depends on mean and quartiles, which in turn can be influenced by how they are estimated [28]. VR index, requiring a more articulated computation algorithm, is more indicated for intra-individual variability, being insensitive to mean sEMG amplitude and data smoothing applied to different waveforms [29]. Neither CQV nor VR indices include the standard deviation in their own definition (Table 1).

Table 1. Different indices to quantify intra-individual sEMG signal variability (elaborated from [27]).

Index	Definition	Parameters
One-dimension CV: it permits comparison of the variability of a data set with a larger and a smaller mean and SD	$CV = \frac{\sqrt{\frac{1}{k}\sum_{i=1}^{k}\sigma_i^2}}{\frac{1}{k}\sum_{i=1}^{k}\lvert\overline{X}_i\rvert}$	k = no. of intervals $^{(*)}$ over a cycle; $\overline{X}_i$ = mean of the sEMG values at the i-th interval calculated over all the cycles; σ_i = standard deviation of the sEMG values calculated over all the cycles.
Two-dimension CV	$CV_i = \frac{\sigma_i}{\overline{X}_i}$	CV at the i-th interval $^{(*)}$. Note that CV is defined as the mean value of CV_i's over the number of intervals in a cycle (k).
Variance Ratio (VR)	$VR = \frac{\sum_{i=1}^{k}\sum_{j=1}^{n}\frac{(X_{ij}-\overline{X}_i)^2}{k(n-1)}}{\sum_{i=1}^{k}\sum_{j=1}^{n}\frac{(X_{ij}-\overline{X})^2}{kn-1}}$ where $\overline{X} = \frac{1}{k}\sum_{i=1}^{k}\overline{X}_i$	k = no. of intervals$^{(*)}$ over the cycle; n = no. of cycles; X_{ij} = sEMG value at the i-th interval for the j-th cycle, $\overline{X}_i$ = mean of sEMG values at the i-th interval over j cycles; $\overline{X}$ = mean of sEMG values.
Coefficient of Quartile Variation (CQV)	$CQV = \frac{(Q_3-Q_1)}{(Q_3+Q_1)}$	Q_1 = 25th percentile, Q_3 = 75th percentile of the n sEMG values at a given interval $^{(*)}$.

$^{(*)}$ Definition of interval depends on the specific study target (e.g., gait analysis, swimming, walking).

As discussed in [30], sEMG-signal amplitude is typically used as a measure of relative force production and it increases with the number, size and firing rate of active motor units. When collecting sEMG, several aspects may affect the measure of sEMG amplitude and frequency, namely the depth of the active motor units, the thickness of the subcutaneous tissues, proximity to the innervation zone and tendons. As such, electrode placement plays a crucial role in sEMG signal quality. Moreover, it is acknowledged that the thickness of the subcutaneous tissue between the surface electrode and active muscles affect the measurement of electromyographic activity. The amount of excess body fat is considered as an internal noise for EMG because it increases the separation between the active muscle fibers and the detection sites [31]. In this work, sEMG signals have been collected from tibialis anterior (TA) and gastrocnemius lateralis (GL) muscles, based on acknowledged guidelines [32,33] for electrodes positioning to maximize the signal-to-noise-ratio. Moreover, obese subjects have been excluded from the study [34]. So, the potentially limiting factor of a small average sEMG value, associated with CV definition, is avoided.

3. Material and Methods

3.1. Participants

A retrospective study was performed, considering sEMG and foot–floor-contact data from 102 volunteer subjects. Volunteers were split into three different groups. H-group was composed of 34 Winters' type I and II hemiplegic children (18 males and 16 females, 6–13 years, age = 7.9 ± 3.0 years, height = 127 ± 18 cm, mass = 27.4 ± 11.0 kg), originally introduced in [4].

C-group was composed of 34 control children (18 males and 16 females, 6–11 years, age = 9.1 ± 1.1, height = 134 ± 9 cm, mass = 32.1 ± 6.9 kg), originally introduced in [35]. A-group was composed of 34 healthy adults (18 males and 16 females, 20–30 years, age = 23.9 ± 1.5 years; height = 174 ± 10 cm; mass = 63.1 ± 12.0 kg), picked up from the populations analyzed in the Movement Analysis Laboratory of Università Politecnica delle Marche, Ancona, Italy and previously introduced in [14] and [36]. Obese subjects were not included in the study. The research was undertaken in compliance with ethical principles of Helsinki Declaration and approved by institutional expert committee. Adult participants signed informed consent prior the beginning of the test. For children, parental consent and child assent were obtained.

3.2. Measurement Chain

Basographic and sEMG signals were acquired and synchronized by means of Step 32 multichannel recording system, (Medical Technology, Turin, Italy, resolution: 12 bit; sampling rate: 2 kHz). Basographich switches (minimum activation force = 3 N), were pasted beneath the heel, the first and the fifth metatarsal heads of each foot, for measuring foot–floor-contact signal. Single differential sEMG probes (Ag/Ag-Cl disk; electrode diameter: 0.4 cm; inter-electrode distance: 0.8 cm; differential amplifier gain: 30 dB; high-pass filter cut-off frequency: 10 Hz; input impedance: 1.5 GΩ; CMRR > 126 dB; input referred noise: 1 Vrms) were placed bilaterally over TA and GL muscles, following acknowledged guidelines [32,33]. Then, subjects walked barefoot back and forth over the floor at preferred speed and pace for at least 2.5 min. Further details about acquisition procedure could be found in [4,35,36].

3.3. Signal Processing

Single gait cycles and the phases within each cycle were assessed from basographic signals following the procedure reported in [37]. Band-pass filtering (20–450 Hz) was applied to raw sEMG signals to remove the baseline drift associated with movement, perspiration, etc., and any DC offset. Further, sEMG signals $x(t)$ were full-wave rectified and then smoothed computing the following RMS formula:

$$RMS = \sqrt{\frac{1}{T}\int_0^T |x(t)|^2 dt} \tag{1}$$

over a sliding window of 50 ms (100 samples). The sliding-window approach allows improving the transitory response and guarantees a better temporal resolution. An example of full-wave rectified (panel A) and RMS (panel B) signals in the same stride is reported in Figures 1 and 2, for GL and TA respectively.

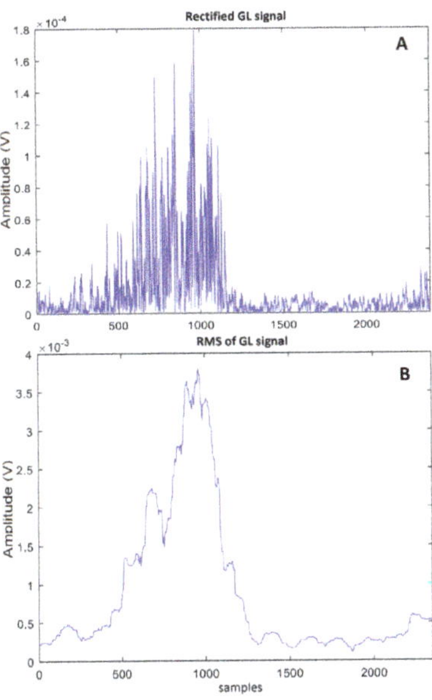

Figure 1. Rectified surface electromyography (sEMG) panel (**A**) and root mean square (RMS) of sEMG signal panel (**B**) for GL in the same representative stride during walking.

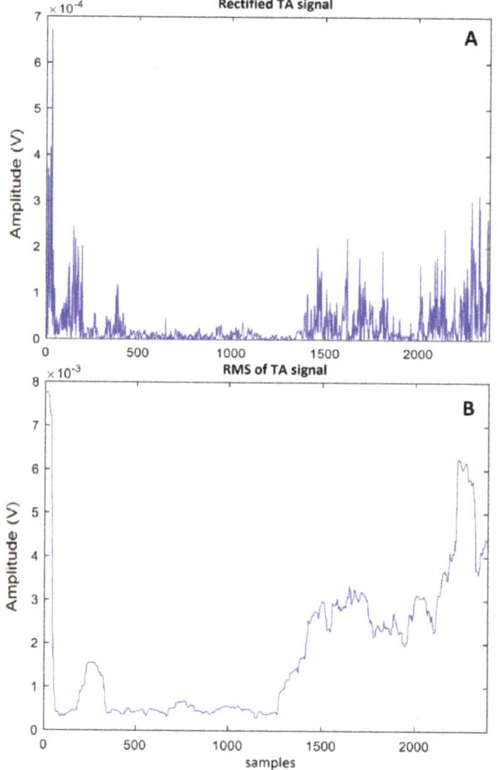

Figure 2. Rectified sEMG panel (**A**) and RMS of sEMG signal panel (**B**) for TA in the same representative stride during walking.

3.4. Variability Index

The CV index is used to measure the variability of muscles rhythmic activation during walking. According to the definition provided in Table 1, the value of this index within a cycle is computed as the ratio of the sEMG signal standard deviation (σ_i) to the mean value ($\overline{X}_i$) in a single *i-th* interval [15,16]:

$$CV_i = \frac{\sigma_i}{\overline{X}_i},\ i = 1 \ldots k \tag{2}$$

As anticipated in Table 1, definition of interval depends on the specific study target (e.g., gait analysis, swimming, walking). The interval considered in the present study is the gait cycle, assessed from the basographic signal. After the evaluation of CV_i index in each single stride, the average over all the k strides of a single walking task gives the global CV. High CV values indicate a large range of variability for a muscle, characterized by periods of contraction and periods of relaxation; lower values identify a more uniform and constant muscle activity.

3.5. Statistics

The Shapiro–Wilk test was used to evaluate the hypothesis that each data vector had a normal distribution. Since all the samples resulted normally distributed, the analysis of variance (ANOVA), followed by multiple comparison test, was used to compare the three groups.

4. Results

In the present study, CV values were assessed over 102 subjects, equally split into the three populations, involving 29,042 strides in total. Figure 3 shows, for every subject of H-group, mean CV values (+SD) over all the available strides for both GL (panel A) and TA (panel B). In the same way, Figures 4 and 5 depict mean CV values (+SD) over all the available strides for both GL (panel A) and TA (panel B), for every subject of C-group and A-group, respectively. Considering 6519 strides in total (a mean value of 192 ± 71 per h-subject), mean CV values (+SD) of 0.71 ± 0.16 for GL and 0.72 ± 0.14 for TA were achieved over H-group. In a total of 9923 strides (a mean value of 292 ± 38 per c-subject), mean CV values of 0.83 ± 0.19 for GL and 0.77 ± 0.12 for TA were computed over C-group. Eventually, in a total of 12,600 strides (a mean value of 371 ± 151 per a-subject), mean CV values of 1.10 ± 0.21 for GL and 0.85 ± 0.11 for TA were obtained over A-group.

A direct comparison among average CV values over the three populations is reported in Figures 6 and 7 for GL and TA, respectively. A statistically significant reduction ($p < 0.05$) of mean CV is detected for GL in H-group compared to both C-group (15% reduction) and A-group (35% reduction). Moreover, the difference observed between C-group and A-group is statistically significant (25% reduction, $p < 0.05$, Figure 6). A significant reduction ($p < 0.05$) of mean CV is detected for TA in H-group compared to A-group (15% reduction). The difference observed between C-group and A-group is statistically significant as well (10% reduction, $p < 0.05$, Figure 7). The 7% reduction of mean CV value detected in H-group compared to C-group is not statistically significant ($p > 0.05$). A direct comparison between mean CV values computed in GL and in TA within the same group was also performed. In A-group, a significant higher mean CV value for GL than for TA was observed (1.10 ± 0.21 vs. and 0.85 ± 0.11, $p < 0.05$). No further significant differences were detected.

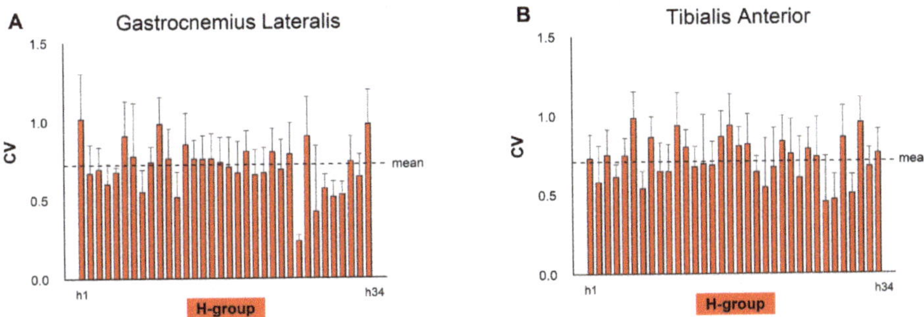

Figure 3. Average coefficient of variation (CV) values (+SD) over all the available strides for GL panel (**A**) and TA panel (**B**) in every subject of H-group. Horizontal dashed line represents the mean value over the H-group.

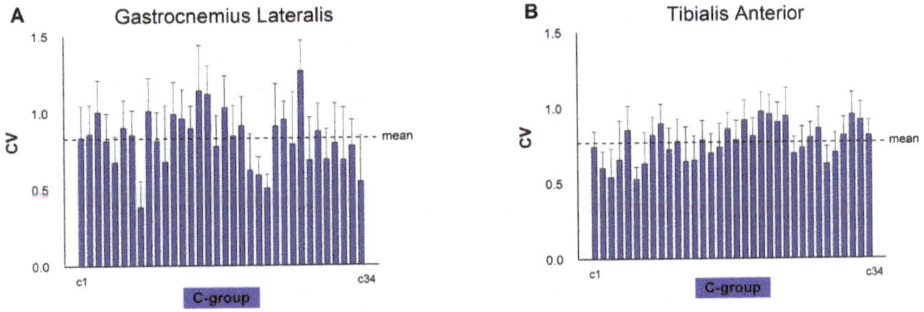

Figure 4. Average CV values (+SD) over all the available strides for GL panel (**A**) and TA panel (**B**) in every subject of C-group. Horizontal dashed line represents the mean value over the C-group.

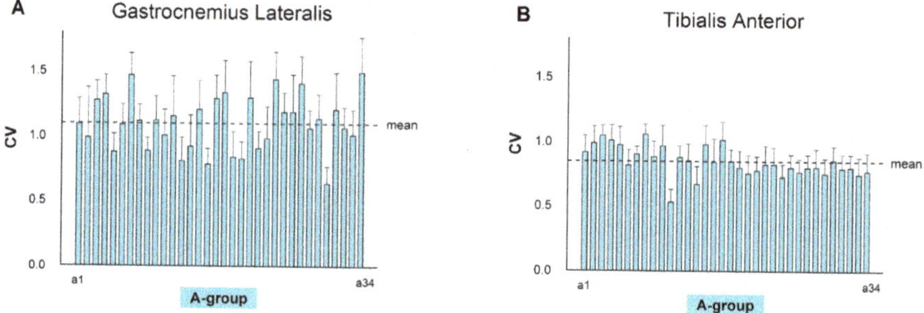

Figure 5. Average CV values (+SD) over all the available strides for GL panel (**A**) and TA panel (**B**) in every subject of A-group. Horizontal dashed line represents the mean value over the A-group.

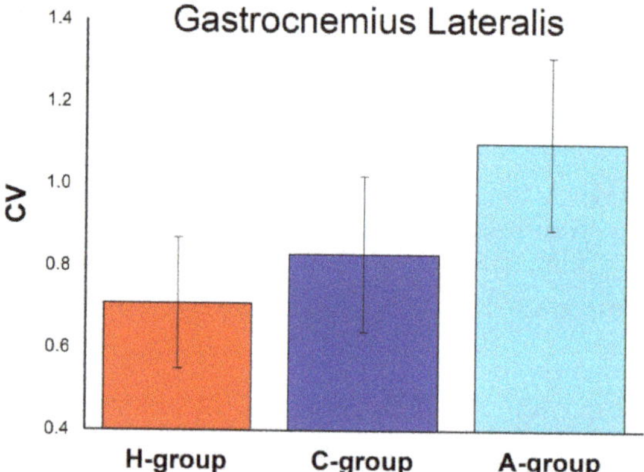

Figure 6. Average CV-GL (± SD) values over H-group (red bar), C-group (blue bar) and A-group (cyan bar). Each mean value is significantly different ($p < 0.05$) from the other two mean values.

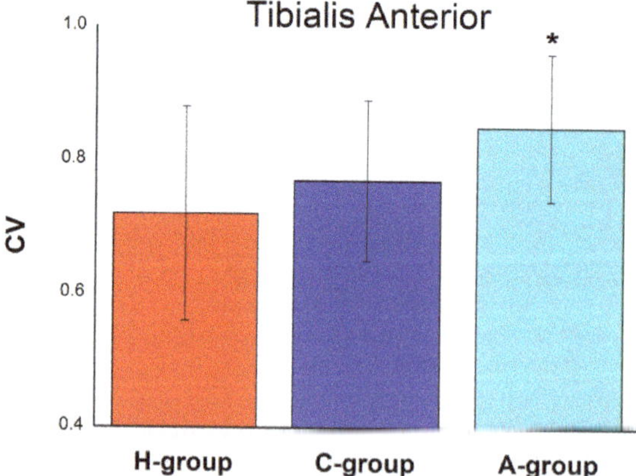

Figure 7. Average CV-TA (± SD) values over H-group (red bar), C-group (blue bar) and A-group (cyan bar). * means significantly different ($p < 0.05$) from the other two mean values.

5. Discussion

Besides the features typically extracted from sEMG signal (RMS, envelope peak, muscle activation timing, median frequency, etc.), some attempts have been recently proposed to consider sEMG-signal variability as a suitable parameter to deepen the interpretation of muscular recruitment by neuromotor system in pathophysiology [38]. Different approaches have been used to quantify this phenomenon [15,39,40]. Nevertheless, a gold standard has not been identified yet. The CV adopted in this work has been proved to be a suitable and easy-to-measure index to assess in different clinical and experimental environments [14–16]. Thus, the goal of the present study is to assess the variability of the sEMG signal acquired over ankle muscles during hemiplegic-children walking by means of CV-index computation. Tibialis anterior and gastrocnemius lateralis are chosen because a large within-cycle variability of sEMG activity in those muscles is reported during hemiplegic-children walking [4,6]. The size of sEMG variability in hemiplegic children is quantified by a direct comparison with CV-based results achieved in a population of age-matched control children and in a further population of able-bodied young adults.

Differently from able-bodied subjects, hemiplegic children are used to hitting the ground in different ways during the same walking, such as by heel, forefoot and flat foot. It has been reported that each one of these contacts would correspond to a different EMG pattern [4]. This is particularly true for those muscles mainly involved in ankle-joint movements, such as GL and TA. Thus, a certain variability of muscle activity is expected, also in mild forms of hemiplegia, such as Winters' type I and II. Results in the present group of hemiplegic children (H-group), indeed, report high (>0.70) mean CV values (±SD) for GL (0.71 ± 0.16) and for TA (0.72 ± 0.14), confirming the above-mentioned reports and previsions. This variability may likely be ascribed to the pathophysiological alternation between sub-phases of gait in which muscles are recruited and sub-phases of gait in which muscles are silent. It could be also observed that sEMG variability is comparable in GL vs. TA, since no significant difference ($p > 0.05$) was detected between mean CV values of the two muscles.

To reliably quantify sEMG variability in hemiplegic children, it is necessary to compare these findings with an age-matched control population, which is represented by C-group in the present study. Alteration of walking in mild hemiplegic children has been widely reported in the literature [2–4,6,8–10]. The classification of hemiplegia proposed by Winters is based only on these differences. Winters' type I children show smaller and less frequent dorsi-flexor recruitment of the ankle in hemiplegic leg, provoking drop foot during swing. This phenomenon is further stressed in Winters' type II, causing a persistence of equinism throughout the gait cycle, often related to a knee hyperextension during stance. This obviously reflects on myoelectric activity of GL and TA. The present study was designed to check if these acknowledged alterations are also associated with a modification of sEMG variability in hemiplegic walking. Results show a decrease of mean CV value in H-group for GL (15%, $p < 0.05$) and TA (7%, $p > 0.05$), compared to C-group (Figures 6 and 7). The CV's own definition indicates that higher values of this index correspond to a more elevated variability of the phenomenon observed. Thus, lower CVs indicate a reduction of intra-subject variability of ankle-muscle activity compared to controls, suggesting that hemiplegic children present a limited capability of adapting their muscle recruitment to the different stimuli met during the walking task, also in the mildest forms of the disease (Winters' type I and II). This consideration is supported by the statistical significance only for GL. Decrease of sEMG variability for TA, indeed, is not statistically significant. This leads to reflect on the meaning of the CV index. As mentioned above, the CV index is not computed directly on sEMG samples, but it is derived from the standard deviation of the signal (Table 1), which is by definition a direct measurement of signal variability. Consequently, it is more informative in the assessment of the differences among different populations than the typical approach based on the statistical comparison among mean values. Thus, in our opinion, the information suggested by the present study could be considered reliable, certainly for dorsi-flexor muscles such as GL, but probably also for plantar-flexor ones such as TA. Moreover, these findings pave the way to further studies which will try and figure out if different results achieved on GL and TA are due to the choice of the index or to the statistical analysis or if this is going to stress a real difference in dorsi-flexor vs. plantar-flexion behavior. A further interesting finding is that H-group presents a larger normalized (to the mean value) range of CV values for both GL (0.33–1.43) and TA (0.63–1.36), compared to C-group (0.61–1.53 for GL and 0.68–1.25 for TA), indicating an increased inter-subject variability of sEMG signals during walking. This result is in line with reported studies indicating that the disorder could affect different patients in different ways [2,4], considering also that the present H-group is composed of both Winters' type I and II children.

A previous research pointed out mean CV values higher than 0.86 for GL in an adult population [16], suggesting that older age could increase sEMG signal variability. Thus, a control group of adult subjects (A-group) was also included in the present analysis, to consider the possible influence of age on CV value. In accordance with the observation reported in [16], CV values in the A-group are significantly higher for GL compared to both C-group and H-group (1.10 ± 0.21 vs. 0.83 ± 0.19 and 0.71 ± 0.16, respectively, $p < 0.05$, Figure 6). This is true also for TA (0.85 ± 0.11 vs. 0.77 ± 0.12 and 0.72 ± 0.14, respectively, $p < 0.05$, Figure 7). Thus, an overall reduction of intra-subject variability is detected in children (hemiplegic and control), suggesting that children are used to adopting a more constant muscular recruitment during walking, with respect to adults. Physiological interpretation of this result may be ascribed to the incomplete maturation of the neuro-motor aspects of walking, acknowledged in school-age children [20]. The CV values reported here in A-group are considerably higher than those shown in [16] for adult people. This is probably due to the difference of gait protocol between the two studies. In the present study, consecutive strides during continuous long-distance gait have been considered. It is reasonable to argue that sEMG patterns may differ and variability could increase, when comparing with signals acquired in single stride during short-distance walking. Moreover, it is acknowledged that a large number of samples are needed to suitably describe the phenomenon of variability of physiological signals [39]. Therefore, the reliability of the present results is strengthened by the numerous strides analyzed here, on average nearly 300 per subject, and 30,000 in total.

The present group of researchers has recently focused its attention on the variability of muscular recruitment in children by means of sEMG analysis, a field where, to our knowledge, only few attempts were carried out. To this aim, different studies were produced, focusing on the assessment of sEMG variability in able-bodied subjects [14,35,36], proposing a new parameter for quantifying sEMG variability [39], looking for novel insights in the maturation of gait [14,36], trying to quantify the asymmetric behavior of muscle recruitment in hemiplegic-children walking [6] and attempting to find a predominant muscle activation pattern able to characterize the different classes of children hemiplegia [8]. However, most of these studies used an advanced signal a processing technique, called statistical gait analysis (SGA), which describes human walking by averaging spatial-temporal and sEMG-based characteristics over numerous strides of the same walking trial. Despite being reliable and robust, SGA is a computationally expensive technique which produces a wide range of results. Thus, the first contribution of the present study is to propose an alternative approach for a suitable assessment of sEMG variability, based on an easy-to-compute and compact index. Table 2 shows the detailed contributions of the present work with respect to each of the abovementioned studies in tabular form. Studies are reported in chronological order. While other studies [5,8,10,11,20,41] investigated muscular recruitment of lower limbs of hemiplegic children during walking, no direct assessment of sEMG variability was reported. Thus, a further contribution of the present study consists in showing the reliability of the CV index in hemiplegic-children walking, in order to also provide information on sEMG variability, besides sEMG amplitude and timing, and all the other typical parameters. A final contribution of the study is the detection of an overall reduction of intra-subject variability of ankle-muscle activity in mild-hemiplegic children compared to controls (children and young adults), suggesting that hemiplegic children present a limited capability of adapting their muscle recruitment to the different stimuli met during walking task. To our knowledge, this information is quantified here for the first time.

Table 2. Detailed contributions of the present work with respect to the state of the art from the same co-authors.

Study	Subject/Patient	EMG Processing	Aim	Results	Contributions of the Present Study
Agostini 2010 [35]	100 able-bodied school-age children	Statistical Gait Analysis (SGA)	To assess variability of muscular timing in numerous strides during walking	Variability was quantified by identifying 5 main activation patterns and their occurrence frequency	Quantification of intra- subject sEMG variability in numerous strides not only in control children, but also in hemiplegic children.
Agostini 2014 [42]	30 hemiplegic children—Winters' type I and II and 100 control children	Statistical Gait Analysis (SGA)	Automatic determination of sEMG patterns of hemiplegic children during gait.	Curtailed activity of tibialis anterior during terminal swing and a lack of activity at loading response in both Winters' class. Class II showed abnormal gastrocnemius activity both at initial contact and in terminal swing	Providing an index for asssessing sEMG variability in order to supply concomitant assessment of sEMG activity and variability
Agostini 2015 [4]	38 hemiplegic children—Winters' type I and II and 100 control children	Statistical Gait Analysis (SGA)	Assessment of variability of muscular timing in numerous strides within each Winters' class during walking	Variability was quantified by identifying 4–5 distinct muscle activation patterns. It cannot be defined a predominant muscle activation pattern for characterizing each specific Winters' class.	(1) Quantification of the decreased intra-subject EMG variability in hemiplegic children compared to both control children and healthy adults (2) Assessment of EMG variability in numerous strides by means of an easy-to-compute index
Di Nardo 2017 [36]	100 able-bodied children and 33 adults	Statistical Gait Analysis (SGA)	Age- and gender-related assessment of EMG variability during walking in control subject to analyze maturation of gait	Increased EMG variability in adult but not in children female, compared to the correspondent male population.	Quantification of the reduced sEMG variability in hemiplegic children compared to both control children and able-bodied adults, providing new insights in maturation of gait and in the effect of hemiplegia on it
Di Nardo2017 [39]	20 able-bodied children and 20 adults	Statistical Gait Analysis (SGA)	To propose the occurrence frequency as a new parameter for assessing sEMG signal variability during walking.	Occurrence frequency is able to provide further information on sEMG variability, besides those supplied by classical temporal sEMG parameters.	Providing an index for asssessing sEMG variability in time domain in order to integrate the information coming from the occurrence frequency

Table 2. *Cont.*

Study	Subject/Patient	EMG Processing	Aim	Results	Contributions of the Present Study
Spinsante 2019 [14]	30 able-bodied children and 30 adults	CV computation	To measure variability of EMG signal in motor development and test the reliability of CV index to this aim	CV index is shown to be able to effectively discriminate pediatric motor capabilities	Extending the reliability of CV index in assessing EMG variability also to hemiplegic-children population
Di Nardo 2019 [6]	16 hemiplegic children—Winters' type I and 100 control children	Statistical Gait Analysis (SGA)	Assessment of variability of muscular timing and asymmetric behavior of muscle recruitment in hemiplegic-children walking	Increased EMG variability in the hemiplegic side due to a reduced activity in terminal swing and a lack of activity at heel-strike of ankle dorsi-flexors.	Testing the reliability in EMG variability assessment of CV index, in a large population including Winters' type I and type II hemiplegic children. This index could be used for an easy-to-compute assessment of hemiplegic asymmetry

6. Conclusions and Future Work

Overall, the present findings provide evidence to support the hypothesis of a decreased intra-subject variability of surface electromyography signal of ankle muscle in hemiplegic children during walking, encouraging future studies to deepen the pathophysiological reasons and modalities associate to this phenomenon. This reduction has been detected compared to both control children and able-bodied adults. Thus, it could probably be ascribed to both young age and the specific disease. Concomitantly, an increased inter-subject variability of sEMG signals was detected during hemiplegic walking, confirming that the disorder could affect different patients in different ways. Furthermore, present findings indicate that CV is a reliable index to evaluate the variability of muscle recruitment in different experimental circumstances and with different clinical goals, such as in adult and pediatric populations, in neurological disorders and for both intra- and inter-subject studies. Including the results obtained from the different indices listed in Table 1, on the set of sEMG measurements collected from the three populations, will be an interesting aspect to investigate in a future development of this study.

It has been shown that the first foot–floor contact of each hemiplegic stride could occur in different ways (with heel, forefoot and flat foot) and that each one of these contacts would correspond to a different EMG pattern. Further research developments could be focused on computing and comparing sEMG variability associated with each one of the different foot–floor contacts, trying to identify which one is more involved in the process of variability decrease. Moreover, it is acknowledged that a single gait cycle can be split in two main gait phases, stance and swing: Stance identifies the full time when the foot is on the ground; swing quantifies the period when the same foot is in the air for limb progression. Assessing sEMG variability separately for stance and swing could be one of the future developments of the present study. Since the CV approach seems to succeed in the quantification of sEMG variability in hemiplegia, further studies could involve other populations affected by neuromuscular disorders, such as cerebrovascular accident, Parkinson's disease and multiple sclerosis.

Author Contributions: Conceptualization, F.D.N. and S.S.; methodology F.D.N. and S.S.; software A.S., A.P., C.P. and S.S.; investigation F.D.N. and S.S.; validation, A.S. and C.P.; resources, S.F., V.A. and M.K.; data curation, F.D.N., A.S. and C.P.; writing—original draft preparation, F.D.N. and S.S., writing—review and editing S.F., V.A., A.P. and M.K.; visualization, F.D.N. and S.S. supervision, S.F. All authors have read and agreed to the published version of the manuscript.

Funding: This research received no external funding.

Conflicts of Interest: The authors declare no conflict of interest.

References

1. Ostensjø, S.; Carlberg, E.B.; Vøllestad, N.K. Motor impairments in young children with cerebral palsy: Relationship to gross motor function and everyday activities. *Dev. Med. Child Neurol.* **2004**, *46*, 580–589. [CrossRef] [PubMed]
2. Winters, T.F.; Gage, J.R.; Hicks, R. Gait patterns in spastic hemiplegiain children and young adults. *J. Bone Joint Surg.* **1987**, *69*, 437–441. [CrossRef] [PubMed]
3. McDowell, B.C.; Kerr, C.; Kelly, C.; Salazar, J.; Cosgrove, A. The validity of an existing gait classification system when applied to a representative population of children with hemiplegia. *Gait Posture* **2008**, *28*, 442–447. [CrossRef] [PubMed]
4. Agostini, V.; Nascimbeni, A.; Gaffuri, A.; Knaflitz, M. Multiple gait patterns within the same Winters class in children with hemiplegic cerebral palsy. *Clin Biomech.* **2015**, *30*, 908–914. [CrossRef]
5. Riad, J.; Haglund-Akerlind, Y.; Miller, F. Power generation in children with spastic hemiplegic cerebral palsy. *Gait Posture* **2008**, *27*, 641–647. [CrossRef]
6. Di Nardo, F.; Strazza, A.; Mengarelli, A.; Cardarelli, S.; Tigrini, A.; Verdini, F.; Nascimbeni, A.; Agostini, V.; Knaflitz, M.; Fioretti, S. EMG-Based Characterization of Walking Asymmetry in Children with Mild Hemiplegic Cerebral Palsy. *Biosensors* **2019**, *9*, 82. [CrossRef]

7. Frigo, C.; Crenna, P. Multichannel SEMG in clinical gait analysis: A review and state-of-the-art. *Clin. Biomech.* **2009**, *24*, 236–245. [CrossRef]
8. Bojanic, D.M.; Petrovacki-Balj, B.D.; Jorgovanovic, N.D.; Ilic, V.R. Quantification of dynamic EMG patterns during gait in children with cerebral palsy. *J. Neurosci. Methods* **2011**, *198*, 325–331. [CrossRef]
9. Galli, M.; Cimolin, V.; Rigoldi, C.; Tenore, N.; Albertini, G. Gait patterns in hemiplegic children with cerebral palsy: Comparison of right and left hemiplegia. *Res. Dev. Disabil.* **2010**, *31*, 1340–1345. [CrossRef]
10. Patikas, D.; Wolf, S.; Döderlein, L. Electromyographic evaluation of the sound and involved side during gait of spastic hemiplegic children with cerebral palsy. *Eur. J. Neurol.* **2005**, *12*, 691–699. [CrossRef]
11. Patikas, D.S.; Wolf, I.; Schuster, W.; Armbrust, P.; Dreher, T.; Döderlein, L. Electromyographic patterns in children with cerebral palsy: Do they change after surgery? *Gait Posture* **2007**, *26*, 362–371. [CrossRef] [PubMed]
12. Turvey, M.; Fitzpatrick, P. Commentary: Development of Perception-Action Systems and General Principles of Pattern Formation. *Child Dev.* **1993**, *64*, 1175–1190. [CrossRef]
13. Hadders-Algra, M. Variation and Variability: Key Words in Human Motor Development. *Phys. Ther.* **2010**, *90*, 1823–1837. [CrossRef] [PubMed]
14. Spinsante, S.; Di Nardo, F.; Strazza, A.; Verdini, F.; Poli, A. A Simple sEMG-Based Measure of Muscular Recruitment Variability during Pediatric Walking. In Proceedings of the 2019 IEEE International Symposium on Medical Measurements and Applications (MeMeA), Istanbul, Turkey, 26–28 June 2019; pp. 1–6.
15. Anders, C.; Wagner, H.; Puta, C.; Grassme, R.; Petrovitch, A.; Scholle, H.C. Trunk muscle activation patterns during walking at different speeds. *J. Electromyogr. Kinesiol.* **2007**, *17*, 245–252. [CrossRef] [PubMed]
16. Bailey, C.; Corona, F.; Pilloni, G.; Porta, M.; Fastame, M.; Hitchcott, P.; Penna, M.; Pau, M.; Côté, J. Sex-dependent and sex-independent muscle activation patterns in adult gait as a function of age. *Exp. Gerontol.* **2018**, *110*, 1–8. [CrossRef]
17. Available online: https://www.myon.ch/ (accessed on 7 August 2020).
18. Available online: https://www.btsbioengineering.com/products/freeemg-surface-emg-semg/ (accessed on 7 August 2020).
19. Agostini, V.; Ghislieri, M.; Rosati, S.; Balestra, G.; Knaflitz, M. Surface electromyography applied to gait analysis: How to improve its impact in clinics? *Front Neurol.* **2020**. [CrossRef]
20. Campanini, I.; Disselhorst-Klug, C.; Rymer, W.Z.; Merletti, R. Surface EMG in Clinical Assessment and Neurorehabilitation: Barriers limiting its use. *Front Neurol.* **2020**. [CrossRef]
21. Donaldson, S.; Donaldson, M.; Snelling, L. SEMG Evaluations: An Overview. *Appl. Psychophysiol Biofeedback* **2003**, *28*, 121–127. [CrossRef]
22. Hermens, H.; Bruggen, T.; Baten, C.; Rutten, W.; Boom, H. The median frequency of the surface EMG power spectrum in relation to motor unit firing and action potential properties. *J. Electromyogr. Kinesiol.* **1992**, *2*, 15–25. [CrossRef]
23. Davies, B.T. A review of "The Co-ordination and Regulation of Movements" By N. Bernstein. *Ergonomics* **1968**, *11*, 95–97. [CrossRef]
24. Guidetti, L.; Rivellini, G.; Figura, F. EMG patterns during running: Intra- and inter-individual variability. *J. Electromyogr. Kinesiol.* **1996**, *6*, 37–48. [CrossRef]
25. Hug, F.; Drouet, J.M.; Champoux, Y.; Couturier, A.; Dorel, S. Interindividual variability of electromyographic patterns and pedal force profiles in trained cyclists. *Eur. J. Appl. Physiol.* **2008**, *104*, 667–678. [CrossRef] [PubMed]
26. Hershler, C.; Milner, M. An optimality criterion for processing electromyographic (EMG) signals relating to human locomotion. *IEEE Trans. Biomed. Eng.* **1978**, *25*, 413–420. [CrossRef] [PubMed]
27. Martens, J.; Daly, D.; Deschamps, K.; Fernandes, R.J.P.; Staes, F. Intra-Individual Variability of Surface Electromyography in Front Crawl Swimming. *PLoS ONE* **2015**, *10*, e0144998. [CrossRef]
28. Raydonal, O.; Marmolejo-Ramos, F. Performance of Some Estimators of Relative Variability. *Front. Appl. Math. Stat.* **2019**, *5*, 43. [CrossRef]
29. Gabel, R.H.; Brand, R.A. The effects of signal conditioning on the statistical analyses of gait EMG. *Electroencephalogr. Clin. Neurophysiol.* **1994**, *93*, 188–201. [CrossRef]
30. Kendell, C.; Lemaire, E.D.; Losier, Y.; Wilson, A.; Chan, A.; Hudgins, B. A novel approach to surface electromyography: An exploratory study of electrode-pair selection based on signal characteristics. *J. NeuroEng. Rehabil.* **2012**, *9*, 24. [CrossRef]

31. Hemingway, M.A.; Biedermann, H.J.; Inglis, J. Electromyographic recordings of paraspinal muscles: Variations related to subcutaneous tissue thickness. *Biofeedback Self Regul.* **1995**, *20*, 39–49. [CrossRef]
32. Agostini, V.; Nascimbeni, A.; Gaffuri, A.; Imazio, P.; Benedetti, M.; Knaflitz, M. Normative EMG activation patterns of school-age children during gait. *Gait Posture* **2010**, *32*, 285–289. [CrossRef]
33. Di Nardo, F.; Laureati, G.; Strazza, A.; Mengarelli, A.; Burattini, L.; Agostini, V.; Nascimbeni, A.; Knaflitz, M.; Fioretti, S. Is child walking conditioned by gender? surface emg patterns in female and male children. *Gait Posture* **2017**, *53*, 254–259. [CrossRef]
34. Available online: https://www.who.int/growthref/who2007_bmi_for_age/en/ (accessed on 17 September 2020).
35. Winter, D.A. *Biomechanics and Motor Control of Human Movement*; John Wiley & Sons: Hoboken, NJ, USA, 2009.
36. Hermens, H.J.; Freriks, B.; Merletti, R. European recommendations for surface electromyography, SENIAM. Enschede (NL). *Roessingh Res. Dev.* **1999**, *8*, 13–54.
37. Agostini, V.; Balestra, G.; Knaflitz, M. Segmentation and classification of gait cycles. *IEEE Trans. Neural. Syst. Rehabil. Eng.* **2014**, *22*, 946–952. [CrossRef] [PubMed]
38. Lord, S.; Howe, T.; Greenland, J.; Simpson, L.; Rochester, L. Gait variability in older adults: A structured review of testing protocol and clinimetric properties. *Gait Posture* **2011**, *34*, 443–450. [CrossRef] [PubMed]
39. Di Nardo, F.; Mengarelli, A.; Strazza, A.; Agostini, V.; Knaflitz, M.; Burattini, L.; Fioretti, S. A new parameter for quantifying the variability of surface electromyographic signals during gait: The occurrence frequency. *J. Electromyogr. Kinesiol.* **2017**, *36*, 25–33. [CrossRef] [PubMed]
40. Tirosh, O.; Sangeux, M.; Wong, M.; Thomason, P.; Graham, H.K. Walking speed effects on the lower limb electromyographic variability of healthy children aged 7–16 years. *J. Electromyogr. Kinesiol.* **2013**, *23*, 1451–1459. [CrossRef]
41. Romkes, J.; Brunner, R. An electromyographic analysis of obligatory (hemiplegic cerebral palsy) and voluntary (normal) unilateral toe-walking. *Gait Posture* **2007**, *26*, 577–586. [CrossRef]
42. Agostini, V.; Knaflitz, M.; Nascimberi, A.; Gaffuri, A. Gait measurements in hemiplegic children: An automatic analysis of foot–floor contact sequences and electromyographic patterns. In Proceedings of the 2014 IEEE International Symposium on Medical Measurements and Applications (MeMeA), Lisbon, Portugal, 11–12 June 2014.

Article

A Systematic Review of Performance Analysis in Rowing Using Inertial Sensors

Matthew TO Worsey, Hugo G Espinosa *, Jonathan B Shepherd and David V Thiel

School of Engineering and Built Environment, Griffith University, Brisbane, QLD 4111, Australia; matthew.worsey@griffithuni.edu.au (M.T.O.W.); j.shepherd@griffith.edu.au (J.B.S.); d.thiel@griffith.edu.au (D.V.T.)

* Correspondence: h.espinosa@griffith.edu.au; Tel.: +61-7-3735-8432

Received: 22 October 2019; Accepted: 5 November 2019; Published: 7 November 2019

Abstract: Sporting organizations such as professional clubs and national sport institutions are constantly seeking novel training methodologies in an attempt to give their athletes a cutting edge. The advent of microelectromechanical systems (MEMS) has facilitated the integration of small, unobtrusive wearable inertial sensors into many coaches' training regimes. There is an emerging trend to use inertial sensors for performance monitoring in rowing; however, the use and selection of the sensor used has not been appropriately reviewed. Previous literature assessed the sampling frequency, position, and fixing of the sensor; however, properties such as the sensor operating ranges, data processing algorithms, and validation technology are left unevaluated. To address this gap, a systematic literature review on rowing performance monitoring using inertial-magnetic sensors was conducted. A total of 36 records were included for review, demonstrating that inertial measurements were predominantly used for measuring stroke quality and the sensors were used to instrument equipment rather than the athlete. The methodology for both selecting and implementing technology appeared ad hoc, with no guidelines for appropriate analysis of the results. This review summarizes a framework of best practice for selecting and implementing inertial sensor technology for monitoring rowing performance. It is envisaged that this review will act as a guide for future research into applying technology to rowing.

Keywords: rowing; technology; inertial sensor; accelerometer; performance; signal processing

1. Introduction

Recent advances in technological developments have enabled the mass production of small, unobtrusive wearable inertial sensors [1]. These sensors can be used to directly monitor an athlete's biomechanics as well as to instrument the equipment an athlete interacts with in a laboratory, training, or competitive setting. Previous studies using wearable inertial sensors for athlete performance analysis show the hindrance of normal movement to be minimal [2]. The miniaturisation of inertial sensors is attributable to microelectromechanical systems (MEMS). MEMS are chip-level devices based on movement of silicon-based arms acting as a mass and spring. The acceleration and rotation can be logged and transmitted [3–5]. MEMS accelerometer technologies include those based on capacitive, piezoelectric, and piezoresistive effects [3]. When inertial sensors are used in rowing, the device must be waterproof. Moreover, if an athlete is being monitored on-water training, then other technologies such as global positioning system (GPS) and on-board video are recommended so that the inertial sensor data can be synchronised.

This review outlines the published literature to assess the applications of inertial sensors in rowing. From this, athletes and coaches have a guide for inertial sensors applications and design method for implementations.

A systematic review evaluating the integration of wearable inertial sensors into a sporting environment for performance monitoring was published in March 2018 [6]. The review captured 286 records and of these 10/286 (3.50%) included on-boat water sports such as rowing and kayaking. These relatively few records either show that there is a gap in using MEMS in performance monitoring in rowing or this is an under researched area. Due to the physically demanding and technical nature of rowing, it is hypothesized that performance monitoring tools would be of great benefit.

Reviews of scientific literature pertinent to rowing have been published; however, to the best of our knowledge, none have focused on the use of technology. Previous literature reviews in rowing have focused on the biomechanical and metabolic factors imperative for a successful rower and the likely injures to rowers. Baudouin and Hawkins [7] looked to bridge the gap between physiological, biomechanics and physical aspects involved in rowing by understanding the interrelationship between the biological and mechanical systems. They propose that the blade force is the only propulsive force counter-acting the drag forces (air drag and hydrodynamic drag acting on the rowing system. It was found that the impact of vertical oscillations of the shell are minimal. The link between blade force and the rower is the oar and this force is transmitted via the oarlock. They suggest that sustainable power is maximized through matching the rigging setup and blade design to the rower's joint torque–velocity characteristics. They concluded that a more comprehensive understanding of force-time profiles are needed so that deficiencies in a rower's biomechanics can be optimised to achieve greater force generation.

Michael et al. [8] reviewed literature surrounding the metabolic demands of kayaking. The scientific literature highlights the high levels of both aerobic power and anaerobic capacity across kayak athletes. They suggested that velocity of the kayak as well as force, power, technique and aerobic fitness are valuable metrics for athlete performance monitoring. Understanding the physiological demands of kayaking is a useful tool for coaches as it helps them make informed decisions about an athlete's suitability for race distances. It also helps to optimise training regimes to improve the performance of specific athletes.

Thornton et al. [9] evaluated published material focused on injuries in rowing. This review was updated in 2016 as rowing specific injury research has increased over the last decade. Key points found from the review were that the largest risk factor for rowing injury were rapid increases in training frequency, intensity and/or volume, appropriate loading in the boat and on a rowing ergometer can reduce the likeliness of overuse injuries, and, finally, there is still a significant demand for well-designed prospective studies focused on rowing injuries. It is evident that an athlete's rowing performance and likelihood of injury can possibly be quantified by metrics obtained via inertial sensors.

The previous literature reviews on rowing do not address the technology and methodologies used in rowing research. Without a framework, ad hoc methodologies concerning the selection and implementation of wearable technologies could reduce the accuracy and validity of rowing sport performance measures [3,10].

2. Materials and Methods

A systematic review of literature was conducted (current as of 24 April 2019) using a methodology based on PRISMA (Preferred Reporting Items for Systematic Reviews and Meta-Analyses) recommendations for completing and reporting the findings of systematic reviews [11]. An electronic database search was completed in total of six relevant scholarly databases (Google Scholar, Web of

Science (core collection), ProQuest, Scopus, Sage Journals and Science Direct) using the keywords identified in Table 1. Exclusion criteria meant manuscripts were only included in the final review if they satisfied the following: It must be a methods-based research article from a scholarly journal (available in English), which contains the use of inertial sensors and have a relevance to human performance monitoring in a rowing setting.

Table 1. Searched databases and associated search terms used, IMU (Inertial Measurement Unit).

Database	Search Terms
Web of Science (core collection)	TS = (Rowing AND sport AND (Inertial sensors OR Accelerometer OR IMU))
Scopus	TITLE-ABS-KEY (("Rowing" AND sport AND ("Inertial Sensors" OR "Accelerometer" OR "IMU))
ProQuest	ALL ("Rowing" AND sport AND ("Inertial Sensors" OR "Accelerometer" OR "IMU))
Science Direct	("Rowing") AND "sport" AND ("Inertial Sensors" OR "Accelerometer" OR "IMU")
Sage Journals	Anywhere ("Rowing") AND anywhere (sport) AND anywhere ("Inertial Sensors" OR "Accelerometer" OR "IMU")
Google Scholar	Rowing, OR Sport, OR IMU, OR Inertial OR Sensor, OR Accelerometer

The included papers were reviewed on the following: (i) the geographical location of where the study was conducted; (ii) properties of the inertial sensor used in the study; (iii) the placement of the inertial sensor in the study; (iv) what algorithms were used for data processing; (v) what performance features were analysed; (vi) study design, and (vii) whether other validated technologies/procedures were implemented to ensure accuracy and validity of the investigation. The record screening process is shown in Figure 1.

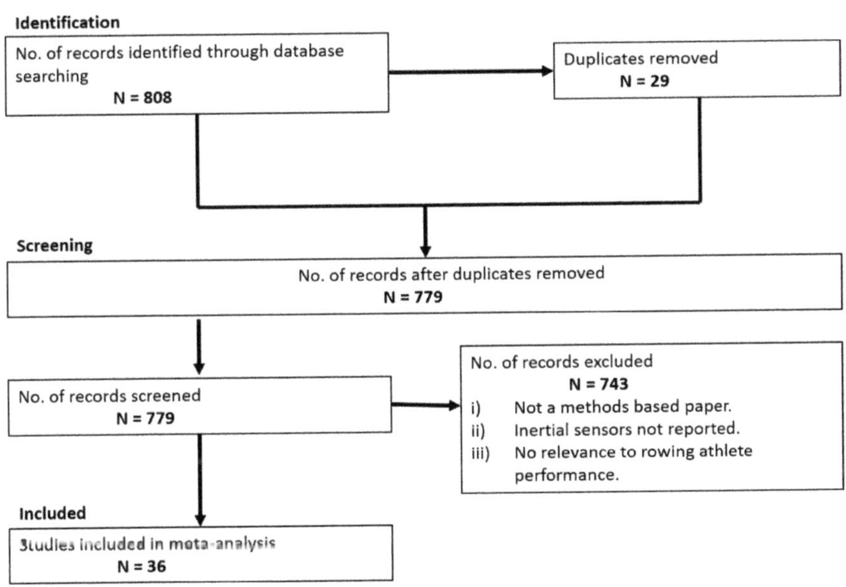

Figure 1. Flow diagram of study selection.

3. Results

3.1. Journals and Years

The papers included in the systematic review were published in several scientific journals. The journals could be divided into three fields of research: Engineering and Technology (78.0%), Sport Science and Medicine (14.0%) and Biomechanics (8.0%) (Figure 2a). The number of papers published each year since 2004 is shown in Figure 2b.

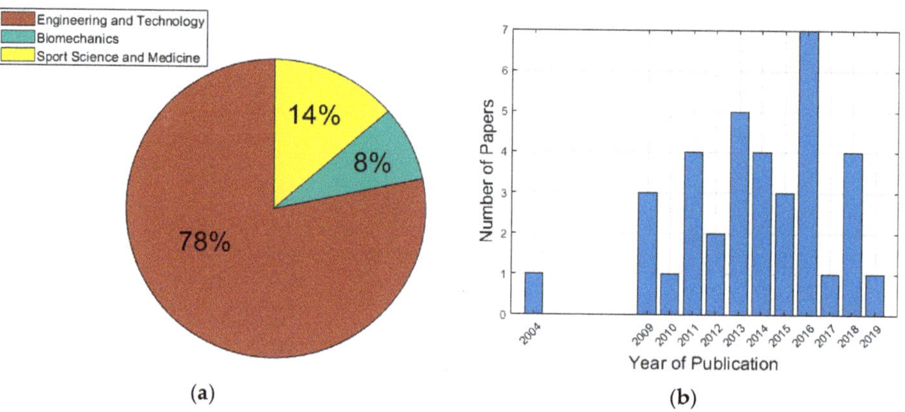

Figure 2. (**a**) journal distribution; (**b**) publication frequency by year. The total number of papers since 2004 is 35.

3.2. Performance Features

An athlete's overall performance in rowing is dependent on many factors. Rowing is an interesting sport to analyse as it has to be considered a system controlled by athlete–equipment interaction. The performance features that were examined in the captured articles were extracted (Table 2). Ultimately, the most important factor for performance in rowing is boat velocity; however, this metric is influenced by various inputs. The features were divided into four main groups: stroke quality, instrumented material metrics, athlete physicality/physiology metrics, and general biofeedback. The stroke quality group consists of metrics such as stroke length, stroke rate, recovery/drive phase ratio, stroke variance, stroke force, and cadence. By instrumenting the equipment used by the athlete, many measures of performance can be observed. Instrumented material metrics included boat position, boat velocity, foot-rest force, boat orientation, oar orientation, and stability. Rowing is an extremely physically demanding sport and requires high skill and therefore the physical and physiological makeup of the athlete is also vital. In this group, measures associated with fatigue, power output, muscle activity, energy output, and also crew synchronicity for team rowing. Finally, general feedback consisted of split times and activity classification.

3.3. Data Processing Algorithms

Various techniques can be used for signal processing of time-series data, for example, frequency filtering of data to remove the effect of noise or drift. Time-series data can also be transferred to the frequency domain using Fourier transforms enabling frequency analysis. Machine learning and deep learning techniques can also be applied for automatic classification of significant events in rowing or human actions, summarized in Table 3. It should be noted that some of the included records used commercialized measurement technology and so limited information about the data processing methods is available.

3.4. Study Design and Hardware

The included records were also reviewed with respect to their study design and the properties of the inertial sensor hardware used. The methodology was evaluated on the number of inertial sensor devices used, the sampling frequency and operating range of the sensors, the location of the device(s), data transmission, the testing environment, and participant selection (Tables 4 and 5). For studies testing the performance of novel hardware or extracting innovative metrics to assess an athlete's performance, then it is important to verify the measurements obtained with a measurements from a 'golden' standard technology. This ensures validity and reproducibility of the measurements obtained (Table 6). In sport and biomechanics research, the golden standard is commonly a multi-camera retro-reflective motion capture system that can track human positions in three-dimensional space. From this data, the acceleration and rotation of the body part can be calculated and compared to inertial sensor data [3]. The competitive setting for rowing is on-water and thus retro-reflective motion capture is not a viable option. It is recommended that measurements are made on a rowing ergometer in a laboratory environment initially. Despite the obvious differences in the biomechanical processes of an athlete when transferring from an ergometer to a scull, it will provide a baseline measurement with golden standard data to validate against.

Table 2. Rowing performance features.

Citation	Sport	Stroke Quality	Instrumented Material Metrics	Athlete Physicality/Physiology Metrics	General Biofeedback
[12]	Rowing	✔	×	✔	×
[13]	Rowing	✔	×	×	×
[14]	Rowing	✔	×	✔	×
[15]	Rowing	×	×	×	×
[16]	Rowing	✔	×	×	✔
[17]	Rowing	×	×	✔	✔
[18]	Rowing	×	✔	×	×
[19]	Rowing	✔	×	×	×
[20]	Rowing	×	×	✔	×
[21]	Rowing	✔	×	×	×
[22]	Rowing	✔	×	×	×
[23]	Rowing	×	×	×	×
[24]	Rowing	×	×	×	×
[25]	Rowing	✔	✔	×	×
[26]	Kayaking	✔	✔	×	×
[27]	Rowing	✔	×	×	×
[28]	Rowing	×	×	✔	×
[29]	Rowing	✔	×	×	×
[30]	Rowing	✔	×	×	×
[31]	Rowing	✔	✔	×	×
[32]	Rowing	✔	×	✔	×
[33]	Rowing	✔	×	×	✔
[34]	Rowing	✔	✔	×	×
[35]	Kayaking	✔	✔	×	×
[36]	Canoeing	✔	✔	×	×
[37]	Kayaking/Canoeing	✔	✔	×	×
[38]	Rowing	✔	×	×	×
[39]	Rowing	×	✔	×	×
[40]	Rowing	×	×	×	×
[41]	Rowing	×	×	×	×
[42]	Canoeing	✔	×	×	×
[43]	Rowing	×	×	✔	×
[44]	Rowing	✔	✔	×	×
[45]	Rowing	×	✔	×	×
[46]	Rowing	✔	✔	×	×
[47]	Rowing	✔	✔	×	×

Table 3. Signal processing algorithms used on data, KNN (K-nearest Neighbours); DTW (Dynamic Time Warping); SVM (Support Vector Machine).

Citation	Sport	Filtering/Windowing Data	Fourier Transform (Frequency Analysis)	Machine/Deep Learning Techniques
[12]	Rowing	×	×	×
[13]	Rowing	×	×	×
[14]	Rowing	✔	×	×
[15]	Rowing	×	×	×
[16]	Rowing	×	×	×
[17]	Rowing	×	✔	KNN
[18]	Rowing	✔	✔	×
[19]	Rowing	×	×	×
[20]	Rowing	×	×	×
[21]	Rowing	×	×	×
[22]	Rowing	✔	×	KNN
[23]	Rowing	×	×	×
[24]	Rowing	×	×	×
[25]	Rowing	×	×	DTW
[26]	Kayaking	✔	×	×
[27]	Rowing	✔	×	×
[28]	Rowing	×	×	×
[29]	Rowing	×	×	×
[30]	Rowing	✔	×	×
[31]	Rowing	×	×	×
[32]	Rowing	×	×	×
[33]	Rowing	×	×	×
[34]	Rowing	✔	×	×
[35]	Kayaking	×	×	×
[36]	Canoeing	×	×	×
[37]	Kayaking/Canoeing	×	×	×
[38]	Rowing	✔	×	×
[39]	Rowing	✔	×	×
[40]	Rowing	✔	×	×
[41]	Rowing	✔	×	×
[42]	Canoeing	✔	×	SVM
[43]	Rowing	✔	×	×
[44]	Rowing	✔	×	×
[45]	Rowing	×	×	×
[46]	Rowing	×	×	×
[47]	Rowing	✔	×	×

Table 4. Properties of inertial sensor instrumentation, Number (#) of devices; OR (Operating Range); RF (Radio Frequency); BT (Bluetooth); ANT (Adaptive network technology); NS (Not Stated); NA (Not Applicable).

Citation	Sport	# of Devices	Accelerometer OR	Gyroscope OR	Magnetometer OR	Sampling Frequency	Transmission
[12]	Rowing	3	NS	NS	NS	NS	RF
[13]	Rowing	2	NS	NS	NS	NS	RF
[14]	Rowing	1x each athlete	±16 g	±2000°/s	±49 Gauss	100 Hz	BT/local
[15]	Rowing	1	NS	NA	NA	100 Hz	NS
[16]	Rowing	3	±6 g	NA	NA	120 Hz	BT
[17]	Rowing	1	±3 g	NA	NA	50 Hz	RF
[18]	Rowing	3	±6 g	NA	NA	83 Hz	RF
[19]	Rowing	1	±10 g	NA	NA	250 Hz	NS
[20]	Rowing	18	±16 g	±2000°/s	±1.9 Gauss	240 Hz wired/ 60 Hz wireless	Real Time
[21]	Rowing	3	NS	±900°/s	NS	NS	Wired
[22]	Rowing	3	±6 g	±2000°/s	NS	200 Hz	RF
[23]	Rowing	2	>±4.2 g	NA	NA	NS	Wireless 802.15.4
[24]	Rowing	1	NS	NA	NA	NS	RF
[25]	Rowing	1	±3 g	NA	NA	50 Hz	NS
[26]	Kayaking	2x each athlete	NS	NA	NA	NS	RF
[27]	Rowing	1	±16 g (scalable)	±2000°/s	NA	200Hz	Local (SD card)
[28]	Rowing	1	±2 g	NA	NA	NS	RF
[29]	Rowing	2	±5 g	NA	NA	NS	NS
[30]	Rowing	1	NS	NS	NS	NS	Wireless
[31]	Rowing	1	NS	NA	NA	25 Hz	RF
[32]	Rowing	1	±8 g (scalable)	NA	NA	≥125 Hz	NS
[33]	Rowing	5	±16 g (scalable)	±2000°/s (scalable)	±12 Gauss	100 Hz	BT
[34]	Rowing	1	±2 g	NA	NA	50 Hz	Wi-Fi
[35]	Kayaking	2	±16 g	±2000°/s	NA	50 Hz	BT
[36]	Canoeing	1	NS	NS	NS	NS	NS
[37]	Kayaking/Canoeing	3	NS	NS	NS	NS	NS
[38]	Rowing	1	±2 g	NA	NA	50 Hz	Wi-Fi
[39]	Rowing	1	NS	NA	NA	100 Hz	Real Time
[40]	Rowing	3	±16 g	±2000°/s	±8 Gauss	100 Hz	RF
[41]	Rowing	3	±3 g	±500°/s	NA	50 Hz	RF
[42]	Canoeing	9	±6 g	±500°/s	NA	100 Hz	RF
[43]	Rowing	14	±16 g	±2000°/s	±1.9 Gauss	60 Hz	Real Time
[44]	Rowing	1	±6 g	±300°/s	NS	50 Hz	NS
[45]	Rowing	2	±6 g	±2000°/s	±4 Gauss	128 Hz	ANT+
[46]	Rowing	1	NS	NS	NS	100 Hz	USB
[47]	Rowing	1	±16 g	±2000°/s	±8 Gauss	100 Hz	BT/local/cloud

Table 5. Sensor placement, Erg (Ergometer); B (Boat).

Citation	Sport	Full Body Model	Forearm/Wrist/Hand	Instrumented Equipment	Back (Upper)	Back (Lower)	Torso/Trunk	Arm(Proximal/Distal)	Leg (Proximal/Distal)	Ear
[12]	Rowing	✔	×	×	×	×	×	×	×	×
[13]	Rowing	×	×	×	✔	×	×	×	✔	×
[14]	Rowing	×	×	Oar	×	×	×	×	×	×
[15]	Rowing	×	×	Erg	×	×	×	×	×	×
[16]	Rowing	×	×	B	×	×	×	×	×	×
[17]	Rowing	×	×	×	×	×	×	×	×	✔
[18]	Rowing	×	×	Oar/**B**	×	×	×	×	×	×
[19]	Rowing	×	×	B	×	×	×	×	×	×
[20]	Rowing	✔	×	×	×	×	×	×	×	×
[21]	Rowing	×	×	Oar/B	×	×	×	×	×	×
[22]	Rowing	×	×	×	✔	✔	×	×	✔	×
[23]	Rowing	×	×	Erg/Oar	×	×	×	×	×	×
[24]	Rowing	×	✔	×	×	×	×	×	×	×
[25]	Rowing	×	×	**B**	×	×	×	×	×	×
[26]	Kayaking	×	×	Oar/B	×	×	×	×	×	×
[27]	Rowing	×	×	B	×	×	×	×	×	×
[28]	Rowing	×	×	Oar	×	×	×	×	×	×
[29]	Rowing	×	×	Oar/B	×	×	×	×	×	×
[30]	Rowing	×	×	B	×	×	×	×	×	×
[31]	Rowing	×	×	B	×	×	×	×	×	×
[32]	Rowing	×	×	B	×	×	×	×	×	×
[33]	Rowing	×	×	×	×	✔	×	✔	×	×
[34]	Rowing	×	×	**B**	×	×	×	×	×	×
[35]	Kayaking	×	×	B	×	×	×	×	×	×
[36]	Canoeing	×	×	Oar	×	×	×	×	×	×
[37]	Kayaking/Canoeing	×	×	Oar/B	×	×	×	×	×	×
[38]	Rowing	×	×	B	×	×	×	×	×	×
[39]	Rowing	×	×	B	×	×	×	×	×	×
[40]	Rowing	×	×	×	×	×	✔	✔	×	×
[41]	Rowing	×	×	×	×	✔	✔	×	✔	×
[42]	Canoeing	×	✔	Oar/B	×	×	✔	✔	✔	×
[43]	Rowing	✔	×	B	×	×	×	×	×	×
[44]	Rowing	×	×	B	×	×	×	×	×	×
[45]	Rowing	×	×	B	×	×	×	×	×	×
[46]	Rowing	×	×	Oar	×	×	×	×	×	×
[47]	Rowing	×	×	Oar	×	×	×	×	×	×

Table 6. Table of inertial sensor validation methods used in the included records.

Citation	Sport	Validation Technology
[12]	Rowing	Optical motion capture
[14]	Rowing	Coach qualitative assessment
[17]	Rowing	Manual feature labelling
[18]	Rowing	Controlled laboratory validation test
[20]	Rowing	Force plates
[25]	Rowing	GPS
[27]	Rowing	GPS
[30]	Rowing	GPS
[32]	Rowing	Navilock-550 ERS
[33]	Rowing	Optical motion capture
[38]	Rowing	GPS
[39]	Rowing	GPS
[40]	Rowing	Optical motion capture
[41]	Rowing	Optical motion capture
[42]	Canoeing	Video Camera
[43]	Rowing	GPS and stroke coach monitor
[44]	Rowing	GPS
[46]	Rowing	Reference measures from rowing simulator
[47]	Rowing	Peach innovations measurement oarlock

4. Discussion

4.1. General Trends

Recent technological developments have made wearable inertial sensors readily available. This has led to sport coaches integrating these devices into their training routines to obtain more measures of an athlete's sport performance in real time [48]. This review highlighted three major research disciplines that are implementing inertial sensors into a rowing setting. These are Biomechanics (n = 3), Sport Science and Medicine (n = 5), and Engineering and Technology (n = 28). The increased availability of inertial sensors is reflected by the rapid increase in the volume of research investigating the use of inertial sensors in rowing since 2004 (Figure 2b). There is a diverse geographic spread of rowing technology research. Countries such as Algeria, Australia, Canada, China, France, Germany, Greece, Italy, Malaysia, Netherlands, New Zealand, Portugal, Slovenia, Spain, Sweden, Switzerland, the United Kingdom, and the United States of America all reported the use of inertial sensors in rowing. The majority of research published originated from Italy (16.7%). Italy won six gold medals at the U23 world championships that may demonstrate the value of integrating sport technology into athlete training programs [49].

4.2. Performance Features

Over half of the included records (24/36—66.7%) used inertial sensors to monitor stroke quality. In rowing, the stroke is the most vital performance indicator and overall performance can be increased by either increasing the propulsive impulse or decreasing the drag impulse within a stroke cycle [7]. The ability to measure the quality of the stroke cycle with an abundance of metrics is thus of high interest to coaches and athletes alike. Stroke rate (cadence) was the most frequently extracted metric surrounding stroke quality (12/24—50.0%) [14,16,21,25,26,33–37]. Athletes can have an optimal stroke rate based on their physicality and thus it is an important metric to monitor. It can also be measured easily with correctly placed inertial sensors. Stroke variability was investigated by seven of the stroke quality concerned records (7/24—29.2%). Stroke by stroke variation in the inertial signals can indicate when athletes are performing well or have faults in their biomechanical processes, stroke variance is also highly investigated in swimming [5]. Three records (3/24—12.5%) addressed the different phases of a stroke, such as the recovery and drive and the ratio of these phases [21,31,32]. The drive/recovery phase ratio is generally used to describe an athlete's rhythm with beginners being advised to aim for

2:1 (drive: recovery). The rhythm of a rower can be directly impacted by increases and decreases in stroke rate; both of these parameters were measured simultaneously by Tessendorf et al. [21]. Stroke length was monitored in three of the included records (4/24—16.7%) [21,26,46,47]. Stroke length is measured in terms of the angle the oar sweeps from catch to finish position. By obtaining measures of stroke length, a coach can tailor training schedules to ensure athletes are entering catch and finish phase of their strokes optimally.

Over a third of the included records (13/36—36.1%) analysed metrics surrounding instrumented rowing materials. Six of the thirteen records investigating instrumented rowing materials (6/13—46.2%) measured boat velocity [25,26,34,35,39,43]. Boat velocity is the most important performance indicator in rowing. GPS is widely accepted as an accurate method to measure boat velocity in rowing; however, the signal is prone to drop outs. Thus, inertial sensors are an appealing method to obtain boat velocity or to work in conjunction with GPS to provide data during periods of drop out. Using inertial sensors to measure velocity can prove to be challenging due to gravitational offsets and sensor drift. Rowing, like swimming, consists of repetitive movements, relatively constant orientation and linear directional movement, providing advantages to the signal processing steps needed to measure velocity [5,50,51].

Five of the instrumented material records (5/13—38.5%) used inertial sensors to measure movement of the oar through the water [18,36,37,46,47]. This can provide coaches and athletes with a visual representation of oar's position through the stroke cycle. This gives potential for a golden standard template of a stroke which enters the different stroke phases at the optimal times to be used as a standard that athletes with certain deficiencies try to replicate. Four of the records (4/13—30.1%) used inertial sensors to monitor the boats position [25,39,44,45], with one of these records investigating the seat position within the boat. Similar to velocity, position is often measured by GPS; nonetheless during drop outs, inertial sensors can assist in interpolation/extrapolation of the GPS data to continuously monitor boat position even when GPS signal drops out. It can also be used to measure stroke efficiency in terms of the distance the boat travels per each stroke. This investigation used an inertial sensor to measure the position of a sliding rowing boat seat. This movement is a direct result of the rower's leg movement and thus is key for optimizing performance. Deficiencies in leg movement of a rower can significantly decrease the velocity of the boat.

Akin to boat position, three of the records (3/13—23.1%) used inertial sensors to monitor the orientation of the boat [18,35,44]. By fusing the sensors within an inertial sensor, typically an accelerometer and gyroscope sometimes accompanied by a magnetometer, an accurate estimation of orientation can be calculated. This is an important metric, as if the boat is deviating too much from its linear path then efficiency is decreased. Thus, athletes can focus on stabilization in their training routines if this is an issue.

Oar stroke force was also an instrumented material metric of interest with three of the records (3/13—23.1%) investigating it. As mentioned earlier, performance in rowing is dependent on increasing propulsive impulse while decrease drag impulse in a stroke cycle [7]. The oar acts as the link between the forces developed by the rower to the blade and creates the propulsive force. Enabling coaches and athletes to measure accurately the oar stroke force is extremely advantageous as strength and conditioning programs can be prescribed to help increase the oar stroke force and in turn overall propulsion of the boat.

Seven of the reviewed papers (7/36—19.4%) investigated rowing athletes' physiological and physicality parameters using inertial sensors. Four of these records (5/7—71.4%) analysed crew synchronicity [12,14,20,23,32]. An elite rower has high special fitness; high coordination, motor control and functional strength [52]. As well as special fitness, team rowers also have to maintain boat stability while staying synchronised with their crew. High synchronicity has been related to increased performance as reflected by the average hull speed [23]. Thus, being able to use inertial sensor data as a measure of synchronicity between rowers can aid coaches in having a more informed understanding of their rowing crew's interactions during training and competition. Armstrong and Nokes [12] investigated synchronization through acceleration signatures and electromyography signals.

This demonstrated the muscle recruitment requirements for different boat positions (stroke, bow). It was also clear to see in the EMG signal the difference between good rowing technique and rowing when 'shooting the slide', which is driving with the legs so the seat leads the back into the drive phase rather than leg and back drive acting as one phase. Two of the records (2/7—28.6%) measured athletes' power output. Estimates of the angular rotation of the oar shaft were obtained using a fitted an accelerometer; the inertial sensors were used in conjunction with force sensors and thus, with the shaft's radius, torque could be derived. Power is the product of torque and angular rotation and can be calculated using an inertial sensor and force sensor. Being able to quantify power output in training and rehabilitation means coaches can monitor their athletes more thoroughly and ensure that their program is achieving efficient results. Atallah et al. [17] (1/7—14.3%) used an earn worn sensor to classify activates and estimate energy output; this was a general study that incorporated rowing as one the activities. However, it does have the possibility to be used purely for rowing.

Three of the reviewed records (3/36—8.3%) provided general biofeedback measures such as tracking the athletes body during rowing, evaluating different methods of providing sonification feedback to rowers and activity classification [16,17,33].

4.3. Algorithms

In order to obtain relevant and insightful metrics from inertial sensor signals, signal processing algorithms have to be used. Presenting raw signal data to a coach or an athlete is sometimes inappropriate as distinct biomechanical events are not distinct; noise can be eradicated from a signal using correctly designed filters. A high volume of the reviewed papers used a windowing/filtering technique during their data analysis (15/36—41.7%) [14,18,22,26,27,30,34,38–44,47].

Of the 15 records, nine (9/15—60.0%) reported the use of a low pass filter [14,18,26,27,30,34,38,39,43]. Seven out of nine of the low pass filters (7/9—77.8%) [14,18,26,27,30,38,43] were used for noise removal. The majority of the noise removal low pass filters were Butterworth (4/7—57.1%) [18,26,27,43] filters ranging from orders of 2–4. The cut off frequencies used for the accelerometer signals ranged from 4 Hz–20 Hz [14,18,26,27,30,43], one record stated the use of a low pass 2nd order Butterworth filter on the gyroscope signal, which had a cut off frequency of 15 Hz [27]. One record used a windowed FIR filter for noise removal but did not state the cut off frequency [38].

Four of the reviewed papers (4/15—26.7%) used sensor fusion algorithms for different purposes [22,39,40,47]. Of these, three-quarters (3/4—75.0%) used sensor fusion computational algorithms to obtain the rower's orientation metrics (e.g., joint angles and oar angles) [22,40,47]. Using inertial sensors to obtain orientation data can produce insightful metrics. The golden standard for these types of biomechanical measures is optical motion capture. By combining anthropometric measurements with the angles obtained by inertial sensors, they can act as a cheaper alternative. Cloud et al. [39] evaluated different sensor fusion (accelerometer and GPS) methods for estimating rowing kinematics such as boat speed and distance travelled. Using the sensor fusion method, the accuracy for boat speed, boat distance travelled and distance per stroke were increased by 44%, 42% and 73%, respectively, when compared to a single channel smartphone GPS.

Machine learning, neural networks and artificial intelligence (AI) algorithms are now frequently applied to sports data for usually time-consuming manual tasks such as feature labelling, classification and future events can be predicted based on existing data. By extracting relevant features in both the time and frequency domains, researchers can apply these algorithms to generate personalized athlete models to further understand their performance. Four of the reviewed records reported the use of these algorithms [17,22,25,42]. Atallah et al. [17] used a KNN model to classify different activities, whereby rowing was one (76.39% success rate). Bosch et al. [22] also used a KNN technique, however, to use inertial sensor signals to distinguish between novice and experienced rowers. The researchers did this by comparing the signals obtained by both experienced and novice rowers to a template generated by an experienced rower. For the most part, the experienced rowers had a closer similarity to the reference rower. The authors concluded that machine learning techniques can distinguish between experienced and novice rowers; however, its hindrance is that it cannot tell the novice rower what their exact deficiency in technique is. Groh et al. [25] used DTW to predict velocity when a GPS signal drops out using inertial sensor data based on the last registered GPS velocity. Wang et al. [42] used SVM classifiers to automatically segment different human motion phases in canoeing; the algorithm was verified by synchronised video footage.

Only two records (2/36—5.6%) analysed data in the frequency domain. Atallah et al. [17] extracted features from the frequency domain to enhance the accuracy of their machine learning model. Llosa et al. [18] used the frequency domain to make more informed decisions about the cut off frequency they used in their signal noise-filter.

4.4. Hardware

Table 4 presents the properties of the inertial sensors used in the 36 reviewed papers. Fourteen of the 36 records (15/36—41.7%) used more than one sensor to make measurements. Seventeen of the 36 records (17/36—47.2%) used an inertial sensor with a built-in accelerometer and gyroscope or magnetometer. Of these seventeen, four (4/17—23.5%) only used an accelerometer and gyroscope and the remaining 13 (13/17—76.5%) also incorporated a magnetometer. The highest reported accelerometer range was ±16 g and the lowest was ±2 g. Compared to our laboratories, previous systematic literature review on the use of inertial sensors in combat sport [3], which reported a maximum and minimum operating range of ±750 g and ±8 g, respectively, these ranges are low. Nonetheless, rowing is a sport that does not typically consist of high impact situations and thus ±16 g and even ±2 g should have a minute risk of sensor saturation.

The measurement devices' sampling frequencies ranged from 25 Hz to 250 Hz, again, compared to combat sport these are low; however, the movement in rowing is far slower compared to strikes in combat sport. Typically, stroke rates in rowing range from 20 strokes per minute (SPM) to 40 SPM (2/3 strokes per second) and thus even a sensor only recording 25 samples per second is going to record the motion with ease. The only concern is that a low sampling rate might only register one sample peaks and troughs, which can lead to underestimates of the true magnitude.

4.5. Study Design

Sensor placement, testing environment and the level of participant(s) in rowing was used to review the records in respect to their study design.

4.5.1. Sensor Placement

Table 5 describes the different sensor placements used across the 36 reviewed manuscripts. The majority of records instrumented a piece of the rowing equipment such as the boat, oars or ergometer with an inertial sensor (27/36—75.0%). Instrumented equipment such as the oars were used to obtain measures such as stroke rate, boat orientation, stroke length, boat velocity, boat position and stability. When the athlete was instrumented, records recorded sensor placements on the forearm, back (upper/lower), torso, arm (proximal/distal), leg (proximal/distal) and ear. Two of the records also used full body inertial sensor systems. When athletes were instrumented, it was often to obtain measures of their body segment orientations throughout a stroke. It is clear, however, that the research community has focused more on instrumenting the rowing equipment with inertial sensors than the athlete and thus there are opportunities for future research investigations doing the latter.

4.5.2. Study Environment

The study environments were classified as either a laboratory/gym or on-water setting. The majority of studies were undertaken on-water (18/36—50.0%) where-by two records had an initial controlled test in the laboratory initially and then used the same methodology on water. Fourteen of the records (14/36—38.9%) conducted their investigations in a controlled laboratory environment, typically on a rowing ergometer; again, two of the records then transitioned the study to an on-water environment.

The remaining four records did not report a study environment. It is promising to see that the majority of the studies have already been conducted in an on-water setting. The advantage of a laboratory setting is that a controlled study is easier to implement, and it offers availability of verification technology such as optical motion capture. Nonetheless, studies conducted in the actual competition environment provide more realistic data for coaching teams and athletes to base training programs around.

4.5.3. Participant Selection

Over half of the included records (19/36—52.8%) recruited participants with experience in rowing. Notable participant recruitments included Olympic and national level rowers [14,16,21,31,46]. It is desirable to recruit participants of a high level within the researched sport as it produces data of higher reproducibility due to the athlete completing biomechanical processes with correct form. Nevertheless, recruiting participants of different levels can strengthen the robustness of machine learning algorithms, improving their accuracy for athletes and the general public. One record recruited novice rowers in order to create a machine learning algorithm which could decipher between experienced and unexperienced rowing techniques within the inertial sensor signals [22].

4.6. Future Recommendations

This review makes it known that stroke quality and instrumented material metrics are the most frequently assessed performance feature in rowing using inertial sensors. This is reflected by the sensor placements extracted from the included records; the majority of records instrumented a piece of rowing equipment. Future research should focus on instrumenting the rowing athlete as well as the equipment and finding an interrelationship between the two. This will help coaches identify faults within the rowing system as a whole.

Machine learning, neural network and AI algorithms are gaining momentum in the sport data field. This systematic review revealed that only four of the 36 records (4/36—11.1%) implemented machine learning algorithms in rowing. Moreover, only two (2/36—5.6%) transformed their data into the frequency domain. Further exploration of these two data processing techniques within the rowing technology field is warranted. Numerous included manuscripts used filtering/windowing processing techniques on their signal data. The majority of filters were used for noise removal and a minority used advanced sensor fusion techniques to calculate values of orientation. Using advanced sensor algorithms such as orientation filters can obtain more insightful metrics about the athlete and/or equipment they are using in terms of Euler Angles and Quaternions, allowing improved information quality and a more detailed assessment of performance.

The highest accelerometer operating range and sample rate reported were ± 16 g and 250 Hz, respectively. Rowing is a low impact sport and subsequently an operating range of ±16 g is acceptable with a low risk as of sensor clipping. MEMS technology is improving rapidly and inertial sensors with sampling frequencies >1 kHz are now readily available with improved sensor resolution. Therefore, it is recommended that future investigations could explore using higher sampling frequencies to improve the quality of the captured information.

5. Operational Guidelines

In similar fashion to the systematic review analyzing the use of inertial sensors in combat sport [3], there is currently no standardization for data collection and analysis procedures in rowing research. Therefore, operational guidelines in the form of a flowchart (Figure 3) and a table (Table 7) are proposed. These guidelines will assist researchers in the selection of technology (device properties, sensor position and validation technology) and data processing algorithms in future rowing technology investigations using inertial sensors. The guideline presented in this review is of the same nature as in the combat sport review as it can be generalized for technology implementation across different sports.

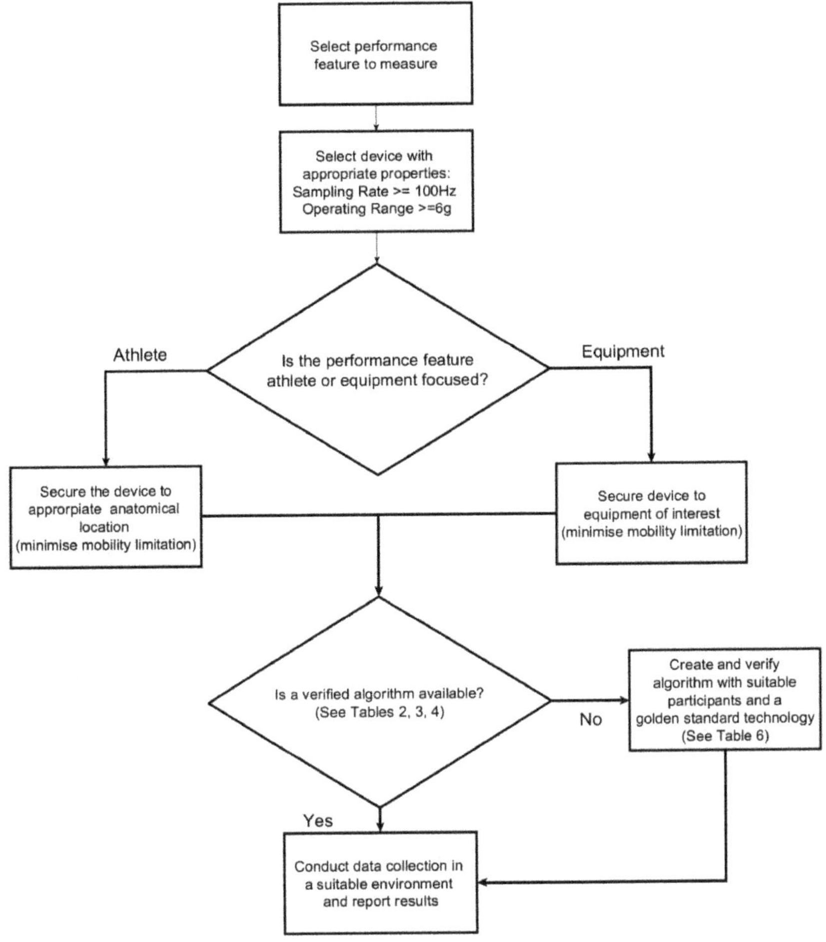

Figure 3. Operational guidelines for methodology design in rowing technology research.

Table 7. Reference guide for technology selection when conducting future research.

Performance Feature	Implementation Complexity	Minimum Hardware Requirements	Minimum Algorithm Implementation
Stroke Quality: Stroke phase ratios/Stroke Length/Stroke Variance	Intermediate to Advanced	Single IMU (on Oar), >6 g Operating range, >100 Hz Sampling Frequency	Threshold peak detection algorithm/advanced orientation algorithm/statistical analysis for variance.
Instrumented Material Metrics: Paddle stroke force/Stability/Boat Position/Boat Velocity/Boat Rotation/Oar Movement.	Advanced	Single IMU (in different positions depending on metric)), >6 g Operating range, >100 Hz Sampling Frequency	Force output estimation equation/Advanced orientation algorithm/Machine learning technique–calculus (w/GPS)/Machine learning technique–calculus (w/GPS)/Advanced orientation algorithm/Advanced Orientation algorithm.
Athlete Physicality/Physiology: Power output/Crew Synchronization//Energy output	Intermediate	Single IMU >6 g Operating range, >100 Hz Sampling Frequency	Power output estimation equation/Correlation analysis/MET estimation calculations
General Biofeedback: Split times/Activity Classification.	Simple to Advanced	Single IMU >6 g Operating range, >100 Hz Sampling Frequency	Threshold peak detection/machine learning technique

Note—the minimum algorithm implementation for each different performance feature is separated by a/and correlates to the first column of the table.

6. Conclusions

Inertial sensors can be used as a performance assessment tool in rowing. In the last decade, research into this field has gained momentum. Inertial sensors were used to measure performance features associated with stroke quality, metrics obtained from instrumented materials, indicators of the athlete's physicality and physiology and also general biofeedback. The review also assessed the properties of hardware that has been used in previous rowing research. Operational guidelines, based on this review, have been created and are proposed to assist future researchers with their methodology design and development. It is suggested that future research starts to incorporate more machine learning/neural network/AI algorithms and starts to focus on instrumenting the athlete rather than the equipment.

Author Contributions: Data retrieval, literature review design, review screening, and manuscript preparation, M.T.O.W.; literature review design, project supervision, and manuscript editing, H.G.E.; project supervision and manuscript editing, J.B.S. and D.V.T.

Funding: This research received no external funding.

Conflicts of Interest: The authors declare no conflict of interest.

References

1. Maenaka, K. MEMS inertial sensors and their applications. In Proceedings of the 2008 5th International Conference on Networked Sensing Systems, Kanazawa, Japan, 17–19 June 2008; IEEE: Kanazawa, Japan, 2008; pp. 71–73.
2. Shepherd, J.B.; Giblin, G.; Pepping, G.-J.; Thiel, D.; Rowlands, D. Development and validation of a single wrist mounted inertial sensor for biomechanical performance analysis of an elite netball shot. *IEEE Sens. Lett.* **2017**, *1*, 1–4. [CrossRef]
3. Worsey, M.; Espinosa, H.; Shepherd, J.; Thiel, D. Inertial sensors for performance analysis in combat sports: A systematic review. *Sports* **2019**, *7*, 28. [CrossRef]
4. Espinosa, H.G.; Lee, J.; James, D.A. The inertial sensor: A base platform for wider adoption in sports science and applications. *J. Fit. Res.* **2015**, *4*, 13–20.
5. Worsey, M.T.O.; Pahl, R.; Thiel, D.V.; Milburn, P.D. A comparison of computational methods to determine intrastroke velocity in swimming using IMUs. *IEEE Sens. Lett.* **2018**, *2*, 1–4. [CrossRef]
6. Camomilla, V.; Bergamini, E.; Fantozzi, S.; Vannozzi, G. Trends supporting the in-field use of wearable inertial sensors for sport performance evaluation: A systematic review. *Sensors* **2018**, *18*, 873. [CrossRef] [PubMed]
7. Baudouin, A. A biomechanical review of factors affecting rowing performance. *Br. J. Sports Med.* **2002**, *36*, 396–402.
8. Michael, J.S.; Rooney, K.B.; Smith, R. The metabolic demands of kayaking: A review. *J. Sports Sci. Med.* **2008**, *7*, 1.
9. Thornton, J.S.; Vinther, A.; Wilson, F.; Lebrun, C.M.; Wilkinson, M.; Di Ciacca, S.R.; Orlando, K.; Smoljanovic, T. Rowing Injuries: An Updated Review. *Sports Med.* **2017**, *47*, 641–661. [CrossRef]
10. Shepherd, J.B.; James, D.A.; Espinosa, H.G.; Thiel, D.V.; Rowlands, D.D. A literature review informing an operational guideline for inertial sensor propulsion measurement in wheelchair court sports. *Sports* **2018**, *6*, 34. [CrossRef]
11. Liberati, A.; Altman, D.G.; Tetzlaff, J.; Mulrow, C.; Gotzsche, P.C.; Ioannidis, J.P.A.; Clarke, M.; Devereaux, P.J.; Kleijnen, J.; Moher, D. The PRISMA statement for reporting systematic reviews and meta-analyses of studies that evaluate healthcare interventions: Explanation and elaboration. *BMJ* **2009**, *339*, b2700. [CrossRef]
12. Armstrong, S.; Nokes, L.D. Sensor node acceleration signatures and electromyography in synchronisation and sequencing analysis in sports: A rowing perspective. *Proc. Inst. Mech. Eng. Part P J. Sports Eng. Technol.* **2016**, *231*, 253–261. [CrossRef]
13. Safiee, M.S.; Ahmad, A.; Tukiran, Z.; Roslan, M.F.; Nadzri, M.M.M. Design and implementation of real-time rowing simulator (ReTRoS) for high-performance and low-injury—A preliminary work. In Proceedings of the 2016 IEEE EMBS Conference on Biomedical Engineering and Sciences (IECBES), Kuala Lumpur, Malaysia, 4–7 December 2016; IEEE: Kuala Lumpur, Malaysia, 2016; pp. 415–419.

14. Avvenuti, M.; Cesarini, D.; Cimino, M. MARS, a multi-agent system for assessing rowers' coordination via motion-based stigmergy. *Sensors* **2013**, *13*, 12218–12243. [CrossRef] [PubMed]
15. Moghaddamnia, S.; Peissig, J.; Schmitz, G.; Effenberg, A.O. A simplified approach for autonomous quality assessment of cyclic movements. In Proceedings of the 2013 18th International Conference on Digital Signal Processing (DSP), Santorini, Greece, 1–3 July 2013; IEEE: Santorini, Greece, 2013; pp. 1–5.
16. Dubus, G. Evaluation of four models for the sonification of elite rowing. *J. Multimodal User Interfaces* **2012**, *5*, 143–156. [CrossRef]
17. Atallah, L.; Leong, J.J.H.; Lo, B.; Yang, G.-Z. Energy expenditure prediction using a miniaturized ear-worn sensor. *Med. Sci. Sports Exerc.* **2011**, *43*, 1369–1377. [CrossRef] [PubMed]
18. Llosa, J.; Vilajosana, I.; Vilajosana, X.; Navarro, N.; Suriñach, E.; Marquès, J. REMOTE, a Wireless Sensor Network Based System to Monitor Rowing Performance. *Sensors* **2009**, *9*, 7069–7082. [CrossRef]
19. James, D.A.; Davey, N.; Rice, T. An accelerometer based sensor platform for insitu elite athlete performance analysis. In Proceedings of the IEEE Sensors, Vienna, Austria, 24–27 October 2004; IEEE: Vienna, Austria, 2004; pp. 1373–1376.
20. Lintmeijer, L.L.; Faber, G.S.; Kruk, H.R.; van Soest, A.J.K.; Hofmijster, M.J. An accurate estimation of the horizontal acceleration of a rower's centre of mass using inertial sensors: A validation. *Eur. J. Sport Sci.* **2018**, *18*, 940–946. [CrossRef]
21. Tessendorf, B.; Gravenhorst, F.; Arnrich, B.; Troster, G. An IMU-based sensor network to continuously monitor rowing technique on the water. In Proceedings of the 2011 Seventh International Conference on Intelligent Sensors, Sensor Networks and Information Processing, Adelaide, Australia, 6–9 December 2011; IEEE: Adelaide, Australia, 2011; pp. 253–258.
22. Bosch, S.; Shoaib, M.; Geerlings, S.; Buit, L.; Meratnia, N.; Havinga, P. Analysis of indoor rowing motion using wearable inertial sensors. In Proceedings of the 10th EAI International Conference on Body Area Networks, Sydney, Australia, 28–30 September 2015; ICST: Sydney, Australia, 2015.
23. Cesarini, D.; Lelli, G.; Avvenuti, M. Are we synchronized? Measure synchrony in team sports using a network of wireless accelerometers. In Proceedings of the 2014 International Workshop on Web Intelligence and Smart Sensing-IWWISS'14, Saint Etienne, France, 1–2 September 2014; ACM Press: Saint Etienne, France, 2014; pp. 1–6.
24. Tukiran, Z.; Ahmad, A. Exploiting LabVIEW FPGA in implementation of real-time sensor data acquisition for rowing monitoring dystem. In *Recent Advances on Soft Computing and Data Mining*; Ghazali, R., Deris, M.M., Nawi, N.M., Abawajy, J.H., Eds.; Springer: Cham, Switzerland, 2018; Volume 700, pp. 272–281. ISBN 978-3-319-72549-9.
25. Groh, B.H.; Reinfelder, S.J.; Streicher, M.N.; Taraben, A.; Eskofier, B.M. Movement prediction in rowing using a Dynamic Time Warping based stroke detection. In Proceedings of the 2014 IEEE Ninth International Conference on Intelligent Sensors, Sensor Networks and Information Processing (ISSNIP), Singapore, 21–24 April 2014; IEEE: Singapore, 2014; pp. 1–6.
26. Gomes, B.B.; Ramos, N.V.; Conceição, F.A.V.; Sanders, R.H.; Vaz, M.A.P.; Vilas-Boas, J.P. Paddling force profiles at different stroke rates in elite sprint kayaking. *J. Appl. Biomech.* **2015**, *31*, 258–263. [CrossRef]
27. Cuijpers, L.S.; Passos, P.J.M.; Murgia, A.; Hoogerheide, A.; Lemmink, K.A.P.M.; de Poel, H.J. Rocking the boat: Does perfect rowing crew synchronization reduce detrimental boat movements? *Scand. J. Med. Sci. Sports* **2017**, *27*, 1697–1704. [CrossRef]
28. Dow, D.; Andrews, R.; Garcia, A.; Dryer, B.; Bonney, S. Rowing training system for on-the-water rehabiliation and sport. In Proceedings of the 8th International Conference on Body Area Networks, Boston, MA, USA, 30 September–2 October 2013; ACM: Boston, MA, USA, 2013.
29. Formaggia, L.; Mola, A.; Parolini, N.; Pischiutta, M. A three-dimensional model for the dynamics and hydrodynamics of rowing boats. *Proc. Inst. Mech. Eng. Part P J. Sports Eng. Technol.* **2009**, *224*, 51–61. [CrossRef]
30. Day, A.H.; Campbell, I.; Clelland, D.; Cichowicz, J. An experimental study of unsteady hydrodynamics of a single scull. *Proc. Inst. Mech. Eng. Part M J. Eng. Marit. Environ.* **2011**, *225*, 282–294. [CrossRef]
31. Kleshnev, V. Boat acceleration, temporal structure of the stroke cycle, and effectiveness in rowing. *Proc. Inst. Mech. Eng. Part P J. Sports Eng. Technol.* **2009**, *224*, 63–74. [CrossRef]
32. Schaffert, N.; Mattes, K. Influence of acoustic feedback on boat speed and crew synchronization in elite junior rowing. *Int. J. Sports Sci. Coach.* **2016**, *11*, 832–845. [CrossRef]

33. Ruffaldi, E.; Peppoloni, L.; Filippeschi, A. Sensor fusion for complex articulated body tracking applied in rowing. *Proc. Inst. Mech. Eng. Part P J. Sports Eng. Technol.* **2015**, *229*, 92–102. [CrossRef]
34. Cesarini, D.; Schaffert, N.; Manganiello, C.; Mattes, K. AccrowLive: A multiplatform telemetry and sonification solution for rowing. *Procedia Eng.* **2014**, *72*, 273–278. [CrossRef]
35. Bifaretti, S.; Bonaiuto, V.; Federici, L.; Gabrieli, M.; Lanotte, N. E-kayak: A Wireless DAQ System for Real Time Performance Analysis. *Procedia Eng.* **2016**, *147*, 776–780. [CrossRef]
36. Morgoch, D.; Galipeau, C.; Tullis, S. Sprint canoe blade hydrodynamics-modeling and on-water measurement. *Procedia Eng.* **2016**, *147*, 299–304. [CrossRef]
37. Umek, A.; Kos, A. Wearable sensors and smart equipment for feedback in watersports. *Procedia Comput. Sci.* **2018**, *129*, 496–502. [CrossRef]
38. Cesarini, D.; Schaffert, N.; Manganiello, C.; Mattes, K.; Avvenuti, M. A Smartphone Based Sonification and Telemetry Platform for On-Water Rowing Training. In Proceedings of the International Conference on Auditory Display, Lodz, Poland, 6–10 July 2013.
39. Cloud, B.; Tarien, B.; Crawford, R.; Moore, J. Adaptive Smartphone-Based Sensor Fusion for Estimating Competitive Rowing Kinematic Metrics. Available online: https://engrxiv.org/nykuh/ (accessed on 10 October 2019).
40. Ben Si Said, K.; Ababou, N.; Ouadahi, N.; Ababou, A. Embedded wireless sensor network for rower motion tracking. In Proceedings of the 2016 8th International Conference on Modelling, Identification and Control (ICMIC), Algiers, Algeria, 15–17 November 2016; IEEE: Algiers, Algeria, 2016; pp. 932–937.
41. King, R.C.; McIlwraith, D.G.; Lo, B.; Pansiot, J.; McGregor, A.H.; Yang, G.-Z. Body sensor networks for monitoring rowing technique. In Proceedings of the 2009 Sixth International Workshop on Wearable and Implantable Body Sensor Networks, Berkeley, CA, USA, 3–5 June 2009; IEEE: Berkeley, CA, USA, 2009; pp. 251–255.
42. Wang, Z.; Wang, J.; Zhao, H.; Yang, N.; Fortino, G. CanoeSense: Monitoring canoe sprint motion using wearable sensors. In Proceedings of the 2016 IEEE International Conference on Systems, Man and Cybernetics (SMC), Budapest, Hungary, 9–12 October 2016; IEEE: Budapest, Hungary, 2016; pp. 644–649.
43. Lintmeijer, L.L.; Hofmijster, M.J.; Schulte Fischedick, G.A.; Zijlstra, P.J.; Van Soest, A.J.K. Improved determination of mechanical power output in rowing: Experimental results. *J. Sports Sci.* **2018**, *36*, 2138–2146. [CrossRef]
44. Mpimis, A.; Gikas, V. Monitoring and evaluation of rowing performance using mobile mapping data. *Arch. Fotogrametrii Kartografii Teledetekcji* **2011**, *22*, 13.
45. Gravenhorst, F.; Thiem, C.; Tessendorf, B.; Adelsberger, R.; Strohrmann, C.; Arnrich, B.; Troster, G. Self-aligning and drift compensated rowing seat position measurment system based on accelerometers and magnetometers. *Proc. Sport.* **2012**, *2012*, 226–231.
46. Gravenhorst, F.; Muaremi, A.; Kottmann, F.; Troster, G.; Sigrist, R.; Gerig, N.; Draper, C. Strap and row: Rowing technique analysis based on inertial measurement units implemented in mobile phones. In Proceedings of the 2014 IEEE Ninth International Conference on Intelligent Sensors, Sensor Networks and Information Processing (ISSNIP), Singapore, 21–24 April 2014; IEEE: Singapore, 2014; pp. 1–6.
47. Gravenhorst, F.; Turner, T.; Draper, C.; Smith, R.M.; Troester, G. Validation of a rowing oar angle measurement system based on an inertial measurement unit. In Proceedings of the 2013 12th IEEE International Conference on Trust, Security and Privacy in Computing and Communications, Melbourne, Australia, 16–18 July 2013; IEEE: Melbourne, Australia, 2013; pp. 1412–1419.
48. Espinosa, H.G.; Shepherd, J.B.; Thiel, D.V.; Worsey, M.T.O. Anytime, anywhere! Inertial sensors monitor sports performance. *IEEE Potentials* **2019**, *38*, 11–16. [CrossRef]
49. Medal Table. Available online: http://www.worldrowing.com/events/2019-world-rowing-under-23-championships/medals (accessed on 15 October 2019).
50. Stamm, A. Velocity and Arm Symmetry Investigations in Freestyle Swimming Using Accelerometry. Ph.D. Thesis, Griffith University, Brisbane, Australia, 2013.
51. Stamm, A.; James, D.A.; Thiel, D.V. Velocity profiling using inertial sensors for freestyle swimming. *Sports Eng.* **2013**, *16*, 1–11. [CrossRef]
52. Kleshnev, V. *Biomechanics of Rowing*; Crowood: Marlborough, UK, 2016; ISBN 978-1-78500-134-5.

www.ingramcontent.com/pod-product-compliance
Lightning Source LLC
LaVergne TN
LVHW070931170726
843515LV00005B/919

* 9 7 8 3 0 3 6 5 0 4 3 8 4 *

MDPI
St. Alban-Anlage 66
4052 Basel
Switzerland
Tel. +41 61 683 77 34
Fax +41 61 302 89 18
www.mdpi.com

Electronics Editorial Office
E-mail: electronics@mdpi.com
www.mdpi.com/journal/electronics